Networked Futures

NETWORKED FUTURES
TRENDS FOR COMMUNICATION SYSTEMS DEVELOPMENT

W.S. Whyte
University of Leeds, UK

JOHN WILEY & SONS LTD
Chichester · New York . Weinheim . Brisbane . Singapore . Toronto

Copyright © 1999 by John Wiley & Sons Ltd
Baffins Lane, Chichester,
West Sussex, PO19 1UD, England

National 01243 779777
International (+44) 1234 779777
e-mail (for orders and customer service enquiries): cs-books@wiley.co.uk

Visit Our Home Page on http://www.wiley.co.uk or
http://www.wiley.com

All Rights Reserved. No part of this publication may be reproduced, stored in a retrieval system, or transmitted, in any form or by any means, electronic, mechanical, photocopying, recording, scanning or otherwise, except under the terms of the Copyright Designs and Patents Act 1988 or under the terms of a licence issued by the Copyright Licensing Agency, 90 Tottenham Court Road, London W1P 9HE, UK, without the permission in writing of the Publisher, with the exception of any material supplied specifically for the purpose of being entered and executed on a computer system, for exclusive use by the purchaser of the publication.

Neither the author nor John Wiley & Sons Ltd accepts any responsibility or liability for loss or damage occasioned to any person or property through using the material, instructions, methods or ideas contained herein, or acting or refraining from acting as a result of such use. The author and Publisher expressly disclaim all implied warranties, including merchantability of fitness for any particular purpose.

Designations used by companies to distinguish their products are often claimed as trademarks. In all instances where John Wiley & Sons Ltd is aware of a claim, the product names appear in initial capital or capital letters. Readers, however, should contact the appropriate companies for more complete information regarding trademarks and registration.

Other Wiley Editorial Offices

John Wiley & Sons, Inc., 605 Third Avenue,
New York, NY 10158-0012, USA

WILEY-VCH Verlag GmbH
Pappelallee 3, D-69469 Weinheim, Germany

Jacaranda Wiley Ltd, 33 Park Road, Milton,
Queensland 4064, Australia

John Wiley & Sons (Canada) Ltd, 22 Worcester Road
Rexdale, Ontario, M9W 1L1, Canada

John Wiley & Sons (Asia) Pte Ltd, 2 Clementi Loop #02-01,
Jin Xing Distripark, Singapore 129809

Library of Congress Cataloging-in-Publication Data
Whyte, Bill.
    Networked futures : trends for communication systems development / W.S. Whyte.
      p.   cm.
    Includes bibliographical references and index.
    ISBN 0-471-98794-8 (alk. paper)
    1. Telecommunication—Technological innovations.
    2. Telecommunication—Forecasting. I. Title.
    TK5101.W453   1999
    621.382—dc21   98-48783
                                                                                        CIP

British Library Cataloguing in Publication Data

A catalogue record for this book is available from the British Library

ISBN 0 471 98794 8

Typeset in 10/12 Times by Footnote Graphics
Printed and bound in Great Britain by Bookcraft (Bath) Ltd
This book is printed on acid-free paper responsibly manufactured from sustainable forestry, in which at least two trees are planted for each one used for paper production.

To Marian, William and Alasdair
August 1998

CONTENTS

Preface		ix
1	Market Needs and Technology Solutions	1
2	New Network Architectures	28
3	Advanced Optical Fibre Networks	83
4	Radical Options for Radio	117
5	Accessing the Network	138
6	Intelligent Management of Networked Information	177
7	Technologies for a Wicked World–Trusted and Trusting Networks	222
8	The Stakeholders	261
9	Applications	300
10	Conclusions	349
11	Appendix: Signal Theory	362
12	Selected Further Reading	376
Index		385

PREFACE

This book is a sober audit of an exciting subject – the Information Superhighway. In it, I try to offer to technical specialists as well as to business professionals, academics and students, the opportunity to develop their own informed opinions on what is hype, what is over-scepticism and what will be possible, as a result of the incredible growth in network speed and computer power we can expect over the next two decades.

Prediction is difficult, obviously, or we would all be millionaires. For more than 30 years I have done my own share of predicting the development and impact of new technology and have seen many colleagues do the same. We have all had very variable levels of success. Often we have been less successful than the non-experts, because of our highly informed prejudices. Sometimes they have failed through lack of depth.

That is why I decided on an across-the-board audit of the emerging technologies and the issues that follow from them, trying to avoid the twin dangers of narrowness and superficiality. The basis of any good audit is the gradual development of a gut feeling that things are or are not on the right track, together with selective drilling-down into the detail to see how well founded are your views. You cannot run a business or design a complex product on the basis of audit results, however good, but you should be able to spot the real strengths, weaknesses and critical issues beneath the superficial.

To give an example: the future of networked systems is dominated by the *fact* that we shall soon have access to vast bandwidth and computer processing power and the *uncertainty* about how we can use it. It is important to keep them separate in one's mind and be able to realise that it will be 'easy' to search the entire on-line databases of the world for a well-specified item of data and retrieve it in a fraction of a second, but extremely difficult to design a computer that can, in any true sense, understand a vocabulary of a few spoken phrases. This is not always made clear by the proselytes of the new millennium.

As technologists we also tend to fall into the trap of assuming that the process of education about the future is one-way, from technology to the rest. This is nonsense; what will happen is more likely to be determined by social change and eternal verities, than by technology alone. People need jobs, are more or less trusting and trustworthy; we employ governments hopefully to fight our corners for us through regulation and progressive taxation; sometimes we demand they protect us from moral outrage; at other times we fear they invade our privacy. I have tried to describe how these issues affect and are affected by our new technology.

In the book, I have given an overview of the major themes, sometimes using self-contained "frames' that, in the absence of hypertext, can be used by those interested in further explanation or detail. In most cases, I have also tried to dissect, with some degree of rigour but without assuming much technical knowledge, significant developments that are illustrative of the state of the art. I have also included an appendix on very basic communication theory, as I believe an appreciation of this greatly enhances one's understanding, although it is not necessary for the rest of the text. (Somewhat to my horror, I realise that many computer engineers have had no training in this area.)

To cover all this in a single text for a disparate readership is not easy – although, I believe, worthwhile – and I am sure I have only been partially successful. Such success as I have had is very much due to the many inspiring opportunities I have had to collect the gossip, the knowledge and the wisdom of my former colleagues at BT Laboratories, my current colleagues at Computer Science in the University of Leeds and my other friends around the world. They have given me their time and their results very freely. I hope I have been able to do justice to their achievements.

Bill Whyte
Leeds, 1998
billw@scs.leeds.ac.uk

1

Market Needs and Technology Solutions

THREE SCENARIOS OF THE FUTURE

- You never travel to work: you live in the Lake District or on a Greek island and your head offices and chief executives, whom you know well although you have never met them, are based in London, New York, Singapore. That's where you and your family shop, too, through all the great stores of the world, without leaving your living room. If you fall ill, you are immediately online to the greatest medical brains that ever lived (and some that never lived). Of course, you still travel, but mainly for pleasure and your route-plan is dynamically optimised, taking into account the conditions of the day, to guide you to a reserved parking space, at a hotel chosen for you by an intelligent computer, that speaks to you as you travel. Always you are in touch with your friends and colleagues – whenever you want to be – and you have at your fingertips (sometimes literally through machines that recognise your gestures) all and only the information you require, in a media-rich format to your taste, that greatly assists your decision-making or your emotional involvement. *You are a beneficiary of a global, intelligent network that provides information, communications, ente you, whenever you want, at prices that are orders of mag those of today.*
- Or imagine you never travelled to work, because you h

to go to. Not very often, at least, because you were part of a marginalised society in a marginalised country that, through its own inability or by self-inflicted blindness, had been unable to adapt to the new realities brought about by technology. Indigenous industry had collapsed and the only way out was to fall back on the scraps provided by a new form of colonialism, a 'Tele-colonialism', that knew no boundaries and against which mere geography was no defence. You have been destroyed by the self-same global network, because you, your company or your country did not, or could not, adopt the new technology fast enough.
- Or virtually nobody works: it is almost all done by machines. They understand your every spoken word and they can find everything you want amid the vast, heterogeneous information sources around the world, presenting it back to you via attractive and realistic personalities that suit your temperament. They do not just know; they act: they carry out medical diagnosis – and surgery – they teach and they entertain. Mankind is close to its Nirvana of purpose.

The first two scenarios are the inescapable alternatives for each of us, according to a number of futurologists. They predict the imminent arrival of a society dominated by ownership of information. In it, the early adopters of new communications and computing technology will triumph over the laggards or the poorer nations because the latter have failed to see the significance of the social changes that will inevitably arise as a consequence of the spectacular growth in power and intelligence of global networks of information. The third scenario, possibly the most frightening and certainly the most extreme, is of a world in which people are more or less redundant, because the machines can do everything, everywhere.

There is, of course, an alternative, equally simplistic scenario, which tries to claim that the future is only an extension of the past, perhaps bigger, but in no way fundamentally different in kind. It goes on to claim that the information society is all 'hype' or, at the very least, a lot further unto the future, and less profound, than the optimists and pessimists predict.

To be truthful, we have never been good at predicting the impact of new technology, or determining whether it will lead to incremental or cataclysmic changes to our lives. In particular, we have always been bad at estimating the rate of change. There is no reason to believe that we shall be any more successful in calculating the impact of the ntinuing improvements in computer and communications technology that underpin the various visions of the 'Information Society', 'Information Superhighway', 'Cyberspace', or whatever we care to call it.

Nevertheless, something is undoubtedly happening and it is the contention of this book that significant, probably dramatic, changes in the way we work and play will occur over the next three decades – one generation – brought about by the ability of new technologies to meet the genuine needs and opportunities of a number of extremely large markets.

This book is intended to flesh out this claim, by looking across the markets and potential applications where intelligent communications can act as a substitute, or a support, to traditional activities, and where it can be used to generate new opportunities. We shall look at technology that will create the infrastructure for this vision and critically assess some of its shortcomings. Arguing that the implications of all of this are on a scale much too great to be allowed to pass without social and political 'interference', we shall also look at some of these issues too. Finally, we shall try to pull together the many and varied strands, to summarise progress, but readers will be expected to make up their own minds as to whether they should be optimists, pessimists or sceptics.

TODAY'S PROBLEMS – TOMORROW'S MARKETS

Major new opportunities do not arise simply through the development of a new technology; what is also required is a market for the applications it makes possible. In the case of information networks, it is the undoubted existence of some very large potential markets that lends most credibility to the bullish claims of the futurologists.

When we look at some of the markets in detail, we see that there is often a common thread that makes a communications solution rather attractive: many, perhaps most, of today's problems are geographical. The world produces enough food, but cannot distribute it. Many more could be cured of disease if medicines and the skills to administer them could reach the point of need. People in some parts of the world work too hard while in others many people can find nothing to do. There are large and difficult distances between those in need of education and those who can deliver it. Even the affluent cities are slowly drowning in pollution and grinding to a halt through traffic congestion. These are all massive failure costs.

Table 1.1 Value/cost for activities with possible communications-based alternatives

Annual UK direct spend on motoring and fares (1992)	£41 bn
UK traffic congestion, per annum	£20–£30 bn
US traffic congestion, per annum	$300 bn
Postal services in the UK, per annum	£6 bn
UK screen-based entertainment, 1991	£4.5 bn
US 'entertainment economy', 1993	$341 bn ('12% of all new jobs')
UK shopping expenditure, per annum	£60 bn (Excluding car sales)
US shopping expenditure, per annum	$750 bn

We can consider some of them to be opportunities, too: for instance, it has been said that 'The entertainment industry is now the driving force for new technology'. Entertainment is a $300 billion-plus economy, largely concerned with the creation, distribution and retailing of information that is set in a richly textured and approachable fashion and, as such, is a potentially voracious consumer of megabytes of data.

Just to take a few examples, we need only to glance at Table 1.1. to see that some very big numbers are involved.

Even if we take a modest view that in under 10% of these examples we can substitute a communications-based alternative, we see that an enormous amount of revenue will be available to fuel the technology revolution. Whilst the installation of fibre optic connection to every household in the UK – that is, the provision of multimegabit bi-directional communication to each and every one of us – has been estimated to cost as much as £10 billion, it becomes a not unreasonable investment, when compared with the potential revenue available.

But, it is reasonable to ask, why should we expect communications-based solutions to take over these markets if they have not done so already? The answer, of course, is that it has been happening and is happening more and more. To give but two examples:

- Only a few years ago, home shopping was called 'mail order': all of the customer orders were posted to the catalogue company. Today, 60% of orders are placed by telephone. Moreover, in the past, many orders were first gathered together by an agent, before being posted off. Now, agents have virtually disappeared. It seems that not only are people aware that telephony provides them with a better, and very possibly cheaper, service, but they are sufficiently confident with the system that they are prepared to use it without the agent.

- In financial services, the figures are even more impressive: in September 1996, First Direct, a UK telebanking operation, had over half a million customers and was growing at nearly 10 000 customers per month. Again the introduction of new technology has led to casualties in the job market: since the introduction of the 'hole-in-the-wall' machine, hundreds of bank branches have been closed and thousands of bank staff have been made redundant.

This 'disintermediation' of the middleman is one of the cost-saving opportunities brought about by information networks and of obvious significance in the labour market. But even if we believe that substitution by communications is inevitable, we need not always think of it as unfortunate. Look again at Table 1.1: in the United Kingdom alone, traffic congestion is estimated to cost £20–£30 billion every year, and in the USA $3000 billion. That is pure failure cost.

If we could find a communications solution to make traffic management more effective, then this would be a massive gain. It is instructive to look at the reasons for our journeys (Table 1.2). The table would suggest that a large number of journeys are discretionary and therefore susceptible to redirection between modes of transport, or even replaceable by other means of communication, and we have every reason to expect that that will happen, for a very simple reason, as Figure 1.1 explains.

We see that telecoms charges have approximately halved every decade, whilst vehicle running costs have remained substantially

Table 1.2 Reasons for travel

Purpose of journey	Proportion of total journeys
Personal business	23%
Shopping	21%
Visiting friends	17%
Commuting	15%
Education	11%
Leisure, entertainment, sport, etc.	9%
Business	4%

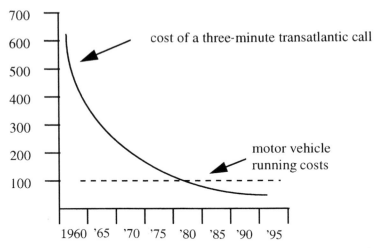

Figure 1.1 Travel and communications – comparative costs (adapted from CEST, 'The way forward', 1993)

constant. Consequently, there will be a strong motive for replacing unnecessary physical travel by a communications alternative. (Consider, for example, the spectacular growth in telephony-based call centres, which handle order taking, complaints, help desk and service desk without customers having to travel or the company requiring high-street presence.)

Looking at the graph, we might conclude that something dramatic must have happened to produce the significant fall in call charges. We would be correct, and the specific reason is technology, the deployment of digital coaxial cables across seas and oceans and the successful launch of telecommunications satellites.

But from the suppliers' point of view, a reduction in price is not to be welcomed unless it is accompanied by a greater increase in purchases. Fortunately for the telecommunications companies, this does, indeed, appear to be the case. Certainly, as Figure 1.2 shows, the last 15 years have produced a dramatic rise in the number of transatlantic voice circuits. There is more than a hint of exponential growth about this curve.

THE GROWTH OF 'NON-VOICE' SERVICES

What is not shown in these graphs is the nature of the traffic being carried. Although the information is given in terms of 'voice circuits',

MARKET NEEDS AND TECHNOLOGY SOLUTIONS

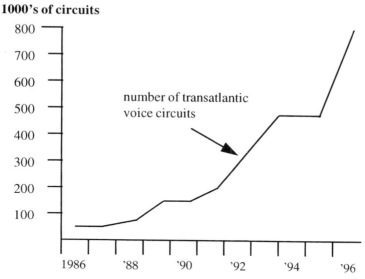

Figure 1.2 Growth in transatlantic voice circuits

which is telecommunications shorthand for a specification regarding the ability to carry an individual telephone call (see the frame 'specification of telecommunications circuits'), the circuits are increasingly likely to carry facsimile or data, or even to be lumped together to carry higher bit rates of data for high-speed computer links or TV channels. For instance, almost 20% of the international traffic through the UK is in the form of facsimile. In fact, non-voice data now accounts for the majority of information carried across telecommunications networks.

THE OUTLOOK FOR PRICES

The curve in Figure 1.2 is for international traffic only and the situation on pricing and volumes is rather complex when we consider a complete basket of services; there are a number of opposing and collaborating factors which will affect the price trend in the future. For instance, there are signs in the case of domestic telephony that price inelasticity and competition are creating downward pressures that are seriously eating into the margins of the telecommunications companies.

SPECIFICATION OF TELECOMMUNICATIONS CIRCUITS

Until a quarter of a century ago, telephony voice signals were almost entirely 'analogue' – the electrical signal on the line was a direct copy of the fluctuations in air pressure that the speaker impressed on the microphone in the telephone handset. The mechanical and acoustical characteristics of the microphone and the earpiece and the characteristics of the line between them, led to a restriction on the frequency range of signals to approximately 300 Hz to 3.4 kHz. The range of amplitude, between the lower limit where signals disappear into background noise, and the upper limit, where the mechanical and electrical components overload and distort the signal, is of the order of 1:1000.

Digital encoding, which has the advantage of cheapness of implementation, significant robustness against problems of line noise and the ability to undergo processing by computers, replaced analogue transmission but retained a specification with broadly the same quality: a standard speech channel now requires that the signal be sampled 8000 times per second and each of the samples 'quantised' into one of 256 discrete levels, which means that each sample can be described by an 8-bit 'word' and the bit-rate per basic speech channel is $8 \times 8 = 64$ Kbit/s.

Even today most domestic telephony is analogue between the local exchanges of the caller and called parties. Thus data signals from computers and fax machines must use modems to send data in analogue form at rates below 20 Kbit/s. Digital conversion of speech or modem signals is carried out at the exchange, where a number of individual voice channels, typically 30, are 'multiplexed' together onto a single coaxial or fibre optic cable, for transmission to the next exchange. These cables carry the speech channels, plus signalling (i.e. dialled digit information) and control information, at a total rate of 2 Mbit/s. On major routes, higher-order multiplexes of a greater number of voice channels are used. R & D activities are examining low-cost options for 2 Mbit/s direct to domestic customers.

Businesses with a requirement for the equivalent of more than a handful of voice circuits are usually provided with a 2 Mbit/s, or higher, digital connection. This can be used as a connection to a PABX on a multiple voice-channel basis for telephony, or, for example, to provide a single 2 Mbit/s pipe for good-quality video-conferencing. Similarly, digital circuits are also provided for computer-to-computer connection, where there is large volume or high-speed data transmission between sites.

Within the last decade there has also been available to business and domestic customers a lower speed, fully digital connection, known as 'basic rate Integrated Services Digital Network', or BR-ISDN. This offers a pair of digital channels, each capable of simultaneously sending and receiving 64 Kbit/s of data, plus a further bidirectional channel of 8 Kbit/s. Each of these channels is independent of the others and can set up calls to other parties, using a fast signalling system that completes the end-to-end connection in a fraction of a second. BR-ISDN offers a high-quality service, suitable for desktop video-conferencing or surveillance (obviously with poorer picture quality than a 2 Mbit/s system, but still adequate for many purposes), and also suitable for computer-to-computer connections at higher rates than can be achieved using modems, or for a range of other tasks such as telemetry and electronic trading.

Finally, mobile networks are rapidly moving towards digital designs that are more tightly specified than their analogue precursors. Digital telephony has, through the necessity to conserve radio bandwidth, to utilise lower bit rates – 16 or even 9 Kbit/s – and more complex methods of encoding the speech.

In this regard it may be significant that the proportion of disposable income spent on domestic telephony in both the US and the UK remains stubbornly well below 10%. (The widely quoted fact that Americans use the telephone much more than the British appears to be more on account of the greater wealth of the former and their free local call policy, than because of a fundamentally different perception of its value.) Experiments with special offers, discounts, etc., have, on the whole, produced depressing results for the network operators, resulting in increased market share but at best zero increase in profitability.

Nevertheless, established operators are harried on all sides by new entrants, frequently creaming off niche markets with high margins, by the competition from cable companies, by the comparatively new entrant of mobile communications and from foreign telecommunications companies now allowed to compete because of the worldwide trend towards liberalisation. They must cut their prices for basic speech services, and cut them dramatically.

We need not be sorry for the telecommunications companies – they are still, on the whole, highly profitable – but what we should realise is that this temporary profitability gives them a window of opportunity in which to tackle the problem. Profitability can be sustained in only two ways:

- increasing revenue by introducing new services
- cutting costs.

In both these cases, solutions that are of sufficient scale to achieve continuing profitability can come about only by introducing technologies which either dramatically reduce the cost per bit per second, or allow new ways of communicating or accessing information.

CURRENT AND FUTURE TELECOMMUNICATIONS 'ARCHITECTURES'

Before considering any possible solutions, it is useful to remind ourselves of the characteristics of the current telecommunications service, its 'architecture', in order better to understand the novel characteristics of new services and the new demands that will be put on the technology. Chapter 2, 'New Network Architectures', will discuss this in detail. For the time being, it is sufficient to remind ourselves that for most of its existence, the telecommunications industry has been based around the provision of one particular type of activity – voice telephony – which has some very specific characteristics, namely the concept of a connection that is:

- bidirectional, usually between two parties only
- between simple terminals (telephones) of similar capability
- with a constant rate of information passing approximately equally in both directions
- at information rates equivalent to that of human speech
- with no redirection to any other party or service during the call
- with no attempt made to interpret the information except for routing ('dialled digits')
- centrally controlled by the network, which is usually provided via one service provider contract.

Many of these characteristics are no longer valid when we come to consider the new markets.

MARKET NEEDS AND TECHNOLOGY SOLUTIONS

NEW SOURCES OF REVENUE – 'ICE' AND INTELLIGENT INFORMATION NETWORKS

A convenient way to summarise the range of new services that could be offered by an intelligent information network is to segment them into three parts, sometimes referred to as the 'ICE' definition. ICE stands for 'information, communication, entertainment'. It is a marketing term, intended to provide a high-level, fairly brutal simplification of the possible set of applications, rather than a detailed analysis of the requirements, but it provides a good summary of how the network needs to change in order to cope with a much wider and diverse set of needs.

Communication Services

Of the three, 'communication' is the nearest to our current understanding of telecommunications. Essentially, it is concerned with 'conversation', in a broad sense, either between humans through voice or e-mail, or in the form of transactions between computers. However, as will be explained in Chapter 3, the requirement for the future will not just be the ability to handle voice and for data. The distinction will largely disappear: the conversations will be multimedia in form, some generating data continuously and some in intermittent bursts, but usually all requiring low end-to-end delay and a guaranteed quality of service. The traffic will have to pass through a number of networks provided by a number of service providers, even within the same country, with a degree of interoperability between their value-added services, such as voice messaging and network-based answering machines. A particularly significant example will be in the progressive integration of fixed and mobile networks, such that one can make calls over fixed lines or radio links, with the combined networks adapting appropriately – for instance, a high late hungry video call originating on a wideband fixed link pressed to provide an adequate picture qu portable videophone.

In order to generate more revenue, telecoms meet the needs of organisations that are incr dispersed and more volatile in structure a needs will be met only by introducing sign into the network, so that, at the simplest le ations will not be burdened with the cumb

from telephone numbers being tied to fixed connections. 'Personal' numbering and mobility will become much more common and organisations will want their communications networks to mirror their organisational structure. The classic example of an application for this type of service is the creation of 'virtual private networks' (see frame) which we already see springing into existence. Fuller integration of voice and computing 'intelligent networks' will be an industry goal.

In order to enable the more advanced features of future communication services, it is likely that terminals will be required that have higher functionality and user-friendliness than simple telephones. Personal computers are obvious candidates for the role of desk-based terminals, with variations on the theme of 'personal organiser' as the

---

VIRTUAL PRIVATE NETWORKS

Consider an enterprise that is distributed across the world, involving a number of people who are on the move and consisting of a consortium of separate companies, working together on range of projects and transfer charging for their services.

At a very simple level, they require to set up multimedia calls and conferences simply by point and click on a participant's name, click on a 'project team' icon to set up a project conference, and so on. Wherever the caller is and wherever the recipient has moved to, the operation is the same; it is the virtual network that translates from name to number and carries out any necessary protocol conversion between the terminals.

At a deeper level, they might require conference membership to be generated automatically from analysis of the structure of project documentation (all documentation is electronic): the creation of the project files generates a communications directory.

Calls between different units might be logged and automatically charged to the appropriate budget. As organisational structures and their members change, the system can be programmed easily and securely to implement these changes.

In dealing with their customers, the organisations may wish to provide 24-hour service by redirecting calls to their service centres ound the world, in a manner that is transparent to the customer, always dials the same number and is charged the same for irrespective of which country hosts the current service

mobile equivalent. Immediately, they raise both an opportunity and a threat to the traditional telco approach, as their inbuilt processing power gives them the opportunity to take a much more active part in the choice and nature of the wide area routing of the connection, challenging the traditional telco concept of an Intelligent Network (see Chapter 3).

It is not just the telecommunications vendors and service providers that will have new challenges to face: the computer companies, in particular, will have to develop better ways of avoiding the quality-of-service problems that beset their distributed systems. Chapter 2 explains some of the serious limitations arising from the democratic, best-efforts routing of traffic within computer networks based on current protocols and practices. The chapter also describes how a compromise position, based on 'Asynchronous Transfer Mode' (ATM), is seen so far as the rather unloved but only solution to the problem.

Information Services

'Information', in the ICE sense, broadly relates to access to databases. Undoubtedly, the best current example of this is the Internet, which has been hugely successful in providing global access to many millions of pages of information. Architecturally, there are significant differences between communication and information networks that are summarised in Figure 1.3.

Networks intended for communications purposes are essentially symmetrical. Information networks, on the other hand, usually handle relatively low data-rates from the user, consisting principally of keyboard or other manual requests for information, and a much larger requirement to send data from the information source to the consumer of the information. Furthermore, the creation and presentation of information is a more costly task than the rea⁻¹ⁱⁿᵍ of it and, consequently, there will be more consumers than cre

The asymmetry of the information flows offers and a dilemma to the traditional telecoms oper Chapter 3, there are possibilities for unidirectio mission, from exchange to customer, over existi there are alternatives available for cable compa broadcast, that will mean that low-cost com emerge. But the telco dilemma is whether t asymmetrical systems or to decide on a m

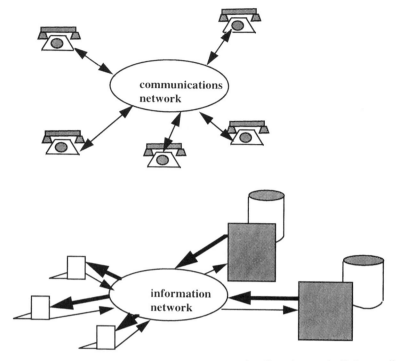

Figure 1.3 Differences between 'communications' and 'information' networks

programme that will provide true high-bit-rate, bidirectional transmission, probably based on optical fibre.

Entertainment Services

If, as we have said, the creation of presentation of information is a more costly task than the reading, hearing or viewing of it, then this is all the more evident in the case of entertainments such as movies or TV programming. Again we envisage a number of database platforms connected to the network, this time holding the entertainment data. However, these platforms will be much more highly 'media rich', with moving images and high-quality sound, requiring much greater data-rates from the supplier. The initial cost of production (of the movie or TV programme that is accessed) will be very high and thus the number of consumers will be orders of magnitude greater than producers. Also, delivery will be predominately into the home and appropriate design of the home terminal becomes an

MARKET NEEDS AND TECHNOLOGY SOLUTIONS 15

issue – is it more likely to be a digital TV set than a home computer? The architecture of the consumer terminal will be considered in Chapter 5.

The major distinguishing feature of entertainment network specifications is in meeting the needs of this asymmetry between production and consumption. Into the marketplace come the major players in the entertainment and consumer electronics industries. The providers will all be fighting to preserve profitability and market share. There will be vigorous technology wars over delivery techniques: telephone wire pairs, cable, optical fibre and terrestrial and satellite broadcasting.

MULTIMEDIA APPLICATIONS OF INFORMATION NETWORKS

It should be clear, from what has already been said, that a general characteristic of the communication, information and entertainment data is its multimedia nature. 'Multimedia communications' is a term defined in a number of ways; for our purpose we define it as follows:

Multimedia communications, through a range of technologies, aspire to provide a rich and immediate environment of image, graphics, sound, text and interaction, which assists in decision-making or emotional involvement.

We can perhaps best set this definition in context by considering some specific applications:

- Training and education – the ability to provide access to remote databases and to interact, face-to-face, with a tutor located elsewhere.
- Distributed work-groups – creating a logically unified project team, with common access to multimedia data (e.g. animated engineering drawings) and video-conferencing.
- Remote experts – providing images from inaccessible locations, e.g. oil rigs, to headquarters experts who can, in return, communicate advice and assembly diagrams.
- Information/sales terminals – located in shops, departure lounges and other public places, to provide information on products or services, perhaps with videotelephone access to a salesperson or customer service.

- Interactive television – entertainment, shopping, education services, accessed via domestic TV sets.
- Computer-based information services; databases of media-rich information accessed by personal computer.

To meet the requirements of multimedia presentation that is fit for its purpose (see the frame below on 'Multimedia requirements') is a demanding task for today's networks, whether they are based on telephone or on computer networks. The former are too expensive for most purposes, but the latter are seriously deficient in quality. There are also, as discussed earlier, incompatible solutions for entertainment versus bidirectional interactive multimedia, that come about because of network inadequacies.

'CONTENT' AND CUSTOMER SERVICE IN A PLURALIST INDUSTRY

The existence of significantly costly 'content' – multimedia information or entertainment accessed over the network – places new requirements for billing, customer registration and all other aspects of customer service. Who, for example, is responsible for handling and rectifying customer complaints when a fault occurs somewhere in the chain between the computer system on which resides a 'movie-on-demand', and the TV set on which it is viewed? Unlike the old days, when one bought a complete telephone service from a single state monopoly, it is unlikely that any one organisation will own everything, from the movie to the point of delivery. Network architectures must be designed on the basis of plurality at every point in the supply chain: multiple information suppliers, multiple information platforms (computer servers), multiple network companies and home and office terminal suppliers. Architectures, both technical and commercial, will have to take this into account.

MULTIMEDIA REQUIREMENTS

The most demanding requirements for multimedia are those needed for transmitting real-time audio and video in both directions. Adequate speech quality can be obtained at a few tens of Kbit/s. The bandwidth for video is higher: tens of Mbit/s are required for high-quality TV, although video compression techniques can produce a quality comparable to decoded satellite broadcasts at 2 Mbit/s. This is also around the lowest acceptable rate for videoconferencing (and we need this 2 Mbit/s for transmitting pictures from *each* conference studio). Smaller size (a window within a personal computer screen, for example) video images that are a little jerky, with poorer definition and which tend to break up if there is too much movement, can be delivered using as little as 64 Kbit/s. A succession of still pictures, refreshed in less than a second, can be sent over the Internet.

Because of the need to conserve bandwidth, all practical coding methods for video, and some for speech, introduce a delay between the capturing of the picture by the TV camera and the generation of the bit stream, plus a further delay between the reception of the data and its conversion to a picture on a screen. Some of this delay is simply due to the time taken for the coding algorithm to work, but some is also due to the fact that the algorithm involves taking the difference between successive picture frames and only transmitting a signal representing the difference between the frames. This latter delay is obviously inherent in the system. Coding and decoding delay is often not a problem in the case of one-way reception (for example, in watching a TV film), provided, of course, the sound is delayed in synchronism, although delay of this nature could be literally fatal if one were carrying out remote surgery, for example. A greater problem occurs when conducting two-way (or multiway) conferencing: the introduction of any delay beyond a few milliseconds in duration creates an artificial, stilted interaction.

In general, the strictest requirement for entertainment multimedia is in the need for sufficient bit-rate to cope with high-quality video and sound; delay is not an issue. In interactive conferencing, the requirement is exactly the opposite: limited bandwidth is acceptable, but long delay is not. Furthermore, entertainment services can afford expensive encoding equipment for the relatively few coding centres, but require cheap decoders for the multiplicity of receivers (e.g. TV sets) that are bought in a very price-sensitive market. The conferencing requirement is, however, obviously for symmetrical coding and decoding.

These differing requirements explain why today's solutions for entertainment and for interactive communication have tended to take different paths in trying to fit their signals into the very limited capabilities of today's networks.

THE CONVERGENCE DEBATE

Much of what we have been discussing above is part of the so-called 'convergence debate', which argues that telecommunications, computing and consumer electronics are, in some senses, the same, and the products and services on offer will converge. This is probably to simplify things far too much: there are similarities between them and their products (as we explain later, there is virtually no difference between a TV set-top box and a personal computer) but there are also significant differences, not least in the appropriateness of each as perceived by the ultimate arbitrator – the user. Some of the argument is cultural; it is interesting to consider telecoms networks versus the 'Internet', purely in terms of their position as retailing operations, selling a service comprising a judicious selection of commoditised products, with a service 'surround'. If we carry the temptation further, perhaps we can consider a venerable concept, the 'Wheel of Retailing', originally proposed by McNair almost exactly 40 yeas ago (Figure 1.4). This proposes a life-cycle for retailing organisations, from start-up to

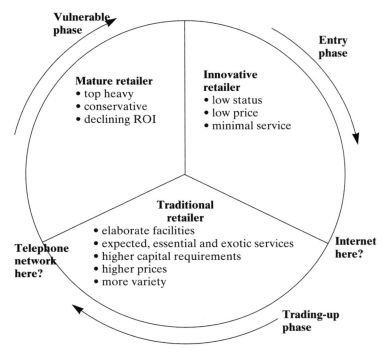

Figure 1.4 The 'Wheel of Retailing'

Table 1.3 Telecoms and Internet 'products'

Telecoms	Internet
staid image	'cyberspace'
expensive	cheap
dominated by quality of service	ad hoc, unspecified connectivity
many differentiated products	few, similar products (http, ftp, e-mail)

maturity, if not ossification. Compare this with what one gets from a telecoms company and from the Internet (Table 1.3).

The appropriateness of the model would appear quite seductive. Convergence of technology will only be one part of the story; there are major cultural shifts required before the technologists can easily work together. Fortunately, there are indications that some progress is being made.

THE SPECIFICATION

We have seen that a large number of new, revenue-generating applications are possible, given an extended capability from communications networks. In particular, what we require is a way of providing:

- high bit-rates at a fraction of today's costs
- guaranteed quality of service from end to end, across heterogeneous equipments and network suppliers
- universal access
- intuitive and helpful interfaces
- intelligent management of information (e.g. finding it, understanding it).

To achieve this aim, we require to look at the role of new technologies.

THE TECHNOLOGY SOLUTIONS

There are a number of possible routes towards the goals, some evolutionary and some revolutionary, but an interesting and encouraging fact is that progress, whatever approach is adopted, rests on the further development of only two fundamental underlying technologies: **optical fibre systems** and **computing power**.

There are relatively safe predictions that can be made in both these areas that can almost guarantee the successful creation of radically different ways of doing business or enjoying domestic services, through communications networks that are orders of magnitude faster and cheaper than those of today and that possess advanced artificial intelligence.

OPTICAL FIBRE SYSTEMS

In Chapter 3, we shall explain how advances in optical fibre technology will allow us to offer services with practically infinite (or, at least, well beyond any foreseen requirement) bit-rates, perhaps even into every domestic premises, at very low costs per bit per second. There has been steady progress in the evolution of fibre in the interexchange network, and designs are beginning to firm up for highly radical, virtually all-optical networks. These will have the potential to provide individual channels of 100 Mbit/s and, by using 'wavelength division multiplexing' from a number of simultaneous laser sources, will provide multiple channels on a single fibre. Wavelength division multiplex also enables a distributed-switching approach, whereby transmission and switching merge and the identity of the 'telephone exchange' becomes blurred. These networks will, through the use of soliton propagation and direct optical amplification using erbium-doped fibre optical amplifiers, have a range which is effectively unlimited.

Optical fibre architectures, not necessarily as radical as this, will open up the possibility of ultra-high-quality images, perhaps even three-dimensional, many of them at a time, transmitted across the world. In addition to this, we shall have the ability to connect the world's computers into these networks and pass data at thousands, if not millions, of megabits per second between each and every one of them. In principle, this combines all of the participating computers into a single supercomputer with a power that we can scarcely imagine.

COMPUTING POWER

Not only that, but each of these individual computers will be subject to the astonishing performance curve that has been consistently adhered to over more than two decades: this curve, 'Moore's Law'; (named

MARKET NEEDS AND TECHNOLOGY SOLUTIONS

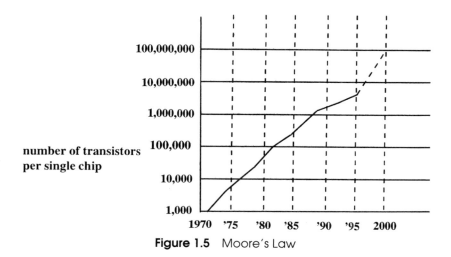

Figure 1.5 Moore's Law

after one of the founders of Intel), describes how the complexity of silicon integrated circuits ('chips') will grow over the years. Taking the basic building block of an integrated circuit to be a 'transistor' – a reasonable assumption, as transistors are the elementary components that logic circuits and amplifiers are constructed from – Moore's Law predicts that each year the number of transistors per chip will double. This prediction, if we allow it to slow occasionally to 18 months, has been astonishingly accurate, as Figure 1.5 shows.

Moore's Law is also approximately true if converted to a statement regarding prices: the computing performance achievable nearly doubles each year, for the same price.

It is generally accepted that the law will be accurate for at least 15 years, implying that computers will achieve almost unimaginable levels of performance by that time.

INTEGRATED INTELLIGENCE AND RADICAL OPTIONS FOR RADIO

Conventional computers, such as the personal computer, are not the only products that can make use of the successful miniaturisation of the electronic circuit. In Chapter 4, we describe how the market for radio systems has exploded in the last few years. Cellular radio products, for example, can stand proudly alongside the Internet as the exponential growers of the 1990s.

This has not come about because someone has discovered missing, unused parts of the radio spectrum – rather the opposite, for the demand for spectrum grows every year. Instead, what has happened is better management of what we have got, made possible by the development of higher-density integrated circuits which control the frequency-agile or multiple-access coded transmitters and receivers, as well as the fixed base station software.

Radio appears to offer a genuinely distinct evolutionary path for communication. This will be predominately, but by no means exclusively, narrowband (i.e. for speech or limited quality pictures). One of the great wildcards of the next decade will be satellite communications, which offer a 'distance-insensitive' cost structure. As such, they are seen by new entrants as a potential total bypass of existing terrestrial networks (although they are careful not to emphasise this, in negotiations with telecommunications regulators and national governments).

ACCESS PLATFORMS

One place where the growth in computing power is already changing the nature of our dialogue with, and through, the network is the access platform. Chapter 4 looks at the possibilities in detail. Already the push-button telephone has begun to be supplanted in many applications by the computer screen, keyboard and mouse. Hardware developments will enhance their suitability for purpose: wall-sized screens will become available, as will very low-power displays, for mobile multimedia applications. The natural platform for information processing will continue to be the personal computer, but it will be challenged by the digital TV set for the provision of entertainment and seductive multimedia. The keyboard, mouse and push-button will still dominate as input devices, but we shall look at radical alternatives such as sensors that can measure the displacement of our hands, as well as other techniques for extending the sense organs, to operate at a distance.

More exotic multimedia telepresence is also examined in Chapter 5. Chapter 6 carries out a critical assessment of the state of the art of speech and gesture recognition, for use as input mechanisms, and speech synthesis for output.

In the case of business applications, the workhorse of information processing and controller of interactive communications will, for the

MARKET NEEDS AND TECHNOLOGY SOLUTIONS

foreseeable future, continue to be the personal computer. Unlike the domestic case, there is extensive local area networking and the speed of these networks needs to increase with the information transfer requirement. As office designs move more and more towards open-plan and the number and types of devices connected to the LAN increases, greater flexibility becomes an issue, and wireless 'connections' employing radio or optics become attractive. Chapter 5 explains some alternatives, as well as the possibility of global 'virtual computers', based on ultra-high-speed connectivity.

THE INFORMATION EXPLOSION

It is widely recognised that effective access to information or to the entire libraries of movies and TV programming does not come about simply because they are stored on a computer that can be accessed online over a communications network. There is too much data for an individual to cope with unaided. Technology is required that can mediate across the vast and heterogeneous data sources, to provide immediate access to selective, relevant, manageable amounts of information, rather than the overwhelming amounts of raw data that already exist today, let alone in the future.

Some of this selection will have to be done by human beings, but automatic selection of information that is new and relevant to individual consumers is an active area of research. Chapter 6, 'Intelligent Management of Distributed Information', looks at this issue in the light of what can be done, manually, to mark up information in a standard way, so that it can be recognised, as well as examining the performance of artificial intelligence when set the same problem. This brings us to a central issue in the feasibility of achieving the bolder visions of the futurologists.

POWER VERSUS INTELLIGENCE

Some very bullish claims are made on the grounds that Moore's Law will provide us with computers so powerful that they can do anything. We need to regard this claim with some scepticism: the *power* of a computing system cannot be directly equated to its *intelligence*. Of course, it is actually very difficult to give a precise definition of intelligence, but, if we limit ourselves to an operational description – the ability of computers to understand speech or to identify objects in moving video, for example – we have to accept that intelligence is not growing in any way as fast as sheer computing power. Computers are still very 'dumb' and the progress in 'artificial intelligence' has been steady rather than spectacular. In many cases, it at least appears that it is not the power that is missing but the underlying theoretical base. A good example of this is in speech processing: it is (relatively) easy to make a computer that can speak in a reasonably natural way, but it is still not possible to make one that understands speech in any fair and honest use of the word 'understand'. Synthesis is one example of an 'easy' problem, which mostly requires lots of data analysis and computation; recognition belongs to the class of 'hard' problems which require something more.

Essentially, the divide between what is 'easy' and what is 'hard' is that the former is achievable because human beings have been able to specify the method of solution in some detail. Chapter 6 explains the basis of the 'frame' problem, the inability of artificial reasoning to place problems within appropriate contexts and to avoid following up every trivial, unrewarding alternative.

That said, it turns out that there is a broad range of activities which are not as difficult as they appear at first sight, and which can be successfully tackled by means of the exponentially increasing power of computing. For example, large-vocabulary machine understanding of dictation is now available, provided the system has been trained on the speaker's voice, and it helps if the subject matter of the dictation has been specified.

Indeed, machines may begin to appear to be intelligent and able to empathise with the users: by combining an intelligent interface with an intelligent database process, it is possible to make interaction with, or across, the network appear like a dialogue with an animated, knowledgeable creature or person, an 'avatar', agent or 'persona', with an image and voice appropriate to the occasion and user. Advances in speech synthesis from raw text can make this feasible, including giving it the characteristics of a famous personality, for instance. Moreover, because we know exactly what sounds are to be made, we also know the position of the lips if the sounds were to be spoken by a person. Therefore, it becomes relatively simple to simulate these lip movements as part of an animation.

Of course, we need to be careful that we recognise that the network personality is not really very intelligent and relies mainly on brute force processing. Much of the next generation of computing will rely on the power of brute force processing rather than the finesse of artificial intelligence.

DISTRIBUTING INTELLIGENCE

One thing we can predict about the nature of artificial intelligence, however rudimentary it might be, is that increasingly it will become 'distributed', with individual computers connected together by the network, to allow processes to run on several machines at once and to communicate with each other in standardised ways. Even the network itself may be optimised by swarms of 'intelligent mobile agents', pieces of computer code that carry out simple tasks and communicate via simple messages, whose collective behaviour is sufficient to control large and complex processes.

Indeed, this interaction between semi-autonomous agents appears highly appropriate as it mirrors the behaviour of the multiparty, virtual organisations that will increasingly make use of intelligent information networks. In this regard, as with other issues of distributed programming, it is important to develop software tools and standards that make it feasible to construct such systems on top of heterogeneous platforms on a multiplicity of separately owned sites. Chapter 2 looks at two distributed architectural approaches – CORBA and TINA – that aim to achieve just that.

TECHNOLOGIES FOR A WICKED WORLD: TRUSTING AND TRUSTED NETWORKS

Whether it is purchasing of goods and services of any kind, or the transmission of valuable corporate data between two geographically separated organisations, there are major issues of security in the protection of the data or in the payment for the services. It is not sufficient to provide a communication path between the parties involved; it is also necessary to make sure that the valuable exchanges of money or secrets can be carried out in a secure way. If networks are to live up to their predicted potential for electronic commerce and global business, they also need to provide security against theft and deception. The technology must be able to withstand potentially highly lucrative attacks from ingenious, unscrupulous and highly skilled criminals, without imposing too much of an additional burden on the millions of innocent people who are potentially customers for the services.

Chapter 7 examines the measures and countermeasures that may create or challenge our trust in networks. There are some techniques that can give physical protection to communication links, but the predominant method of protection will be by means of encryption.

To what extent can machines be enabled to recognise us? Here again we run into the distinction between intelligence and power: machine intelligence capable of recognising voices, fingerprints, handwriting and so on, is progressing slowly, but miniaturisation of integrated circuits means it is now possible to encrypt this data, the user's photograph and other personal data on a microprocessor so small it can be sealed within a plastic credit card, along with the encryption hardware required to make it secure.

STAKEHOLDERS

It is not just criminals who can have a major impact on the acceptability or otherwise of the adoption of the information network; there are a large number of other influential groups (see Figure 1.6) which have, or feel they have, a stake in something with such a potential to change the way we live and work. Chapter 8 examines some of these stakeholder issues and their likely impact on technology.

Regulation is an area that can require significant consideration by anyone planning a technical architecture. There always exists some form of regulatory boundary between controlled and uncontrolled parts of the network and planners must decide on which side of this boundary they place the services. Convergence creates extra problems: regulators must work for fair competition, not just throughout individual domains of computing, telecommunications and consumer electronics, but also across these different areas; it is not always clear where the technical issues end and the commercial ones begin.

People are not always willing to leave it to the regulator to sort things out; there are many who will seek redress at law for defamation or infringement of copyright, there is a minefield of issues relating to decency and appropriateness across a global communication system that must behave acceptably within national boundaries, there are people with disabilities and disadvantages of location who have rights in society, and there are rights to privacy that clash with the fight against crime. Finally, the concept of the nation-state itself is

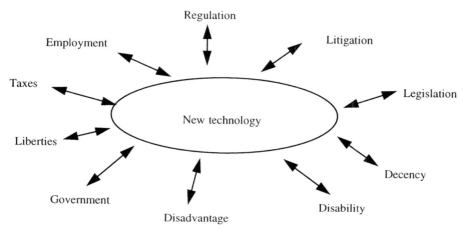

Figure 1.6 Impacts of technology

called into question by technologies that do not recognise boundaries. All these create major technical design issues which must be considered at the outset of designing the information society.

WINNING APPLICATIONS AND TECHNOLOGIES

At the beginning of this chapter, we suggested that there were a large number of very large markets that might gain from development of intelligent information networks. In Chapter 9, we look at a number of them. We shall see that they employ a combination of the technologies that are described in the intervening chapters: some require very large bandwidth; others a significant amount of artificial intelligence. They cover a wide field, and there is virtually no part of our work, play, education or health that they do not impinge on; virtually all of them are in prototype, at least, today. One prophecy that we can make, with a reasonable degree of certainty, is that we have failed to spot some of the most significant applications. However, there is enough business to be had, even from the ones that we have identified, to create a convincing business case to further the arguments of the futurologists who say that the arrival of global information networks is imminent and immensely important to us all. In Chapter 10, we try, probably rashly, to put some sort of timescale to it and to suggest how the technologies will move, in order to make things happen.

2
New Network Architectures

DEFINING AN ARCHITECTURE

Architecture is primarily about how function (a place to live, a place of worship, a way of conveying speech over long distances) is represented in a consistent overall form rather than as a listing of its detailed components – the bricks, the windows, the electromechanical relays. But it is nevertheless determined to a significant degree by the characteristics of these basic materials: a church built of wood is unlikely to have exactly the overall form of one built of brick, just as a telephone network based on radio links is unlikely to mirror exactly one that uses optical fibre.

There are always considerable arguments about what constitutes an architecture, even whether the term is a legitimate one to use for anything other than real buildings but if we simply use it as a way of making it easier to understand the most significant features of the networks and their attachments, emphasising difficulties and requirements to meet changing circumstances, then we can justify its use.

Figure 2.1 shows what is probably the simplest, useful architectural view of tomorrow's information networks: terminals of different types cooperate across a plurality of networks to provide a range of services to support. The architecture is 'layered'; that is, the figure distinguishes between

- the basic physical means of carrying the signal – radio, optical fibre, copper wire

NEW NETWORK ARCHITECTURES

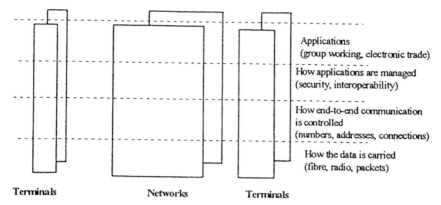

Figure 2.1 A simple architecture for information networks

- the way the connection is controlled, independent of the physical means – for example, is there something equivalent to a setting-up period preceding something like a telephone 'call' followed by a closing-down procedure?
- the services – making a computer and a telephone interoperate, for example
- the innumerable applications that run on top of all the above.

There are other architectural diagrams that use more or fewer levels, define them rather differently or even use extra dimensions at each layer. Architectural models are useful, conceptually, but seldom are real systems defined from a purely architectural viewpoint. In the discussions that follow, we necessarily jump between layers in the architecture – usually without warning – and that is why we have chosen what might be considered to be a lowest common denominator that can be conveniently retained at the back of the mind.

INTEGRATION

The greatest challenge facing the coming generations of networks is that of integration: integration of the old legacy with the new, and integration of the locally optimum solutions for radio, computing, telephony and consumer electronics, because we wish our networks to carry out an integration of services (Figure 2.2):

1. We want to provide a truly global service
2. for people on the fixed (wired) network

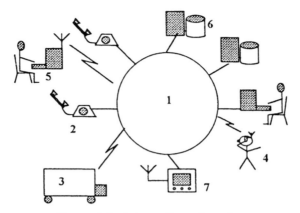

Figure 2.2 Integration of services

3. to talk to each other and people on the move
4. with personal communicators,
5. including mobile workstations/personal computers
6. and those on a fixed network,
7. and to allow computers to pass information to each other and to humans,
8. and to involve entertainment and other consumer products,
9. and also to enable many to communicate to many
10. in a rich multimedia environment
11. security and reliably at economic cost,

THE HISTORICAL LEGACIES

If the networks of the next decade are to meet the potential markets for communication of information and entertainment, providing them globally and on the move, with rich multimedia environments, there will have to be some major changes in the 'network architecture'. Today, there are two competing architectural viewpoints, that of telecommunications and that of computer networks. Neither alone is sufficient and both will have to evolve to meet their individual needs. These needs will converge, and we shall explore how this convergence may come about, both by fairly orderly evolution and by the revolution brought about by the explosion of bandwidth made available by optical fibre. First, we need to look at the architectural heritage.

TELEPHONY TRANSMISSION ARCHITECTURE – THE 'TAPERED STAR'

Telephony naturally developed a centralised architecture as soon as it became desirable to connect a large range of people by means of individual pairs of wires (the only technological option at the time). If A wishes to connect to any of n people, then n possible connections are required; if B now wants to connect to any of the other, $(n-1)$ further connections are required (one less because B is already connected to A). For all the n customers to be connected to each other requires $n + (n-1) + (n-2) \ldots + 2 + 1$ total connections; that is, a number of the order of n^2. If all telephones in Great Britain were to be connected together by a dedicated pair f wires, we should require about one billion pairs of wires going into our houses. Thus, early on, it became obvious that the system could work only if the individual wires were taken to a central point, a telephone 'exchange', where they could be selectively connected when required.

Reducing the number of direct connection points from several million customers to several thousand telephone exchanges meant that more money could be spent on each individual inter-exchange route. The cables could be 'better', in terms of resistance to electrical noise or by having lower attenuation of the signal per mile. Most significantly, by deploying coaxial cable and modulation schemes such as time division multiplex (see frame, page 33), several hundred separate telephone calls could be carried on one cable. The telephone network therefore resembles a 'tapered star': each individual customer is provided with one pair of wires (per telephone 'line') making the connection to the local exchange or, in some cases, a simpler 'remote concentrator'; at the local exchange or concentrator all the customers' connections are electronically multiplexed onto the better, inter-exchange cables; long-distance calls may be switched onto even higher-capacity circuits that run between main exchanges (Figure 2.3).

Almost all of the inter-exchange circuits of first-world networks are digital, using time domain multiplexing. Originally this was on a link-by-link basis, but progressively since the 1980s inter-exchange networks have been moving towards a centralised, synchronised digital highway, based heavily on optical fibre for the 'higher order multiplex' parts – those where the greater number of individual channels are combined together – with all the timing provided from a central, master clock. The basic building blocks are 49.5 Mbit/s data

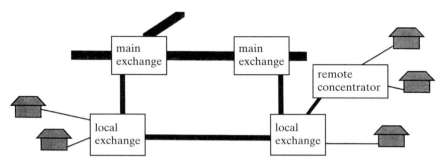

Figure 2.3 The tapered star

streams, multiplexed together into a number of blocks or 'hierarchies' (Table 2.1).

These are the bit-rates available for carrying user-data. Added on to these figures is control information for the network. It might conceivably be possible to negotiate a contract with a telecommunications company to purchase exclusive use of one of the lower-rate blocks, but it is not really intended that users will want to take up the offer, because the price would be astronomical. Users are expected to be able to afford much lower bit-rates: many small organisations are 'happy' with tens of Kbit/s and even large businesses, with big switchboards or lots of data to transport, generally make do with a few 2 Mbit/s connections.

Note that, whatever amount of data the user passes over the network, the charging is based on the fixed bit-rate capability of the circuit.

In general, the connection between the customer and the local exchange is the bottleneck that sets the maximum available bandwidth

Table 2.1 The synchronous digital hierarchy

SDH hierarchy, Mbit/s
49.536
148.608
445.824
594.432
891.648
1188.864
1783.296
2377.728

or, equivalently, bit-rate, of the call. Many of the connections from the local exchange to business customers and virtually all of the local to domestic connections are analogue, with each pair of wires limited to 3.4 kHz bandwidth, allowing modem signals of around 20 Kbit/s. In this regard, we should note that telephony provided by cable TV companies does not really differ from that available from the traditional telecoms companies; the telephony is not, in general, supplied over the coaxial cable used for the TV signal; instead, it is provided on conventional telecoms cables. The emergence of the Integrated Services Digital Network (ISDN) provides a facility of dial-up, digital connections, with data-rates from 64 Kbit/s upwards. Each 64 Kbit/s channel is priced around the same as a voice telephone call. Businesses can get economies of scale by ordering 2 Mbit/s connections (for example from a PBX or a router for computer data) and private circuits can be provided for point-to-point permanent connections.

PACKET NETWORKS

In passing, we also mention a completely different type of network, run by telecoms companies, that of a 'packet network'. These were set up three decades ago to deal with the relatively slow data-rates involved when people communicated, intermittently, with remote computers. Users were reluctant to pay the price of a continuous voice

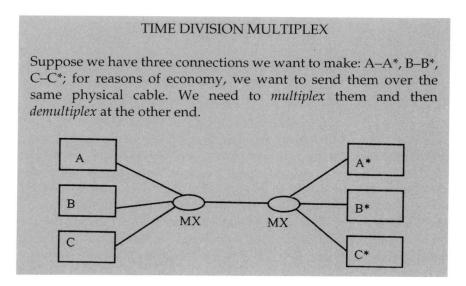

TIME DIVISION MULTIPLEX

Suppose we have three connections we want to make: A–A*, B–B*, C–C*; for reasons of economy, we want to send them over the same physical cable. We need to *multiplex* them and then *demultiplex* at the other end.

One effective way to do this is by *time division multiplex (TDM)*. Consider three bit-streams, A, B, C, each occurring at the same rate, R. We can shorten the duration of each of the bits (leaving spaces between them) and then interleave them into a single bit-stream which has a rate of $3R$:

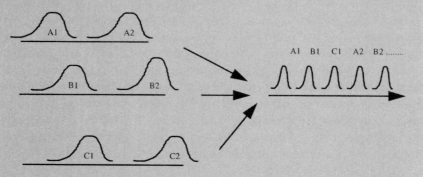

Obviously, the cable which carries them must be able to carry signals at rate $3R$. Slightly less obviously, the multiplexors must be capable of operating at rate $3R$ if they are to be able to interleave the individual signals and then fan out the composite one. Current and planned telecommunications systems, whether optical or not, maintain a hierarchy of multiplexing, where customers can, or will soon be able to, enter at a variety of rates, but the lowest individual circuits start at a basic rate of 64 Kbit/s and multiplexed (together with control and signalling data) into groups of 30, at 2 Mbit/s, which in turn, for longer distances, can be aggregated into 155 Mbit/s, or even 2–4 Gbit/s, at which point most multiplexing equipment runs out of steam.

connection and, as an alternative, they were offered a network which passed blocks of data ('packets') through a series of relays that stored data and then passed it on through network connections that were completely separate from the 'normal' telephone network. By storing the data and delaying it until a path was free, it was possible to make the system cheaper than a voice network – and the delay also ensured that it could not be used as a way of bypassing the telephone network! The service specification promised delivery of the packet in the correct order, with minimal chance of losing a packet, but without a tight specification on network delay. Charging could be based, at least in part, on the number of bits transmitted, rather than on 'call

NEW NETWORK ARCHITECTURES

duration'. Although packet switching never remained close to the centre of mainstream telecommunications thinking over its first 30 years, it promises to have great significance for the next, as we shall see in later sections.

TELEPHONY CALL CONTROL ARCHITECTURE – CENTRAL INTELLIGENCE

As well as transmitting calls across the network, it is necessary to route them between the correct points. Just as with the transmission architecture, call control follows a centralised model.

In the early days of telephony, when the telephone handset was lifted, a switch was closed resulting in a lamp lighting at the operator's position. The operator then asked the caller for the number wanted and worked out the best way to switch the call across the network (Figure 2.4).

Figure 2.4 Manual operator switching

Thus, the network 'intelligence' resided with the operator, in the centre of the network. The terminals, that is, the telephones, were completely 'dumb'. With the advent of automatic dialling, the telephone terminal was able to generate simple control messages, by sending a series of open circuit/closed circuit pulses to the centralised, automatic control in the exchange, where these pulses operated a stepping switch, which moved round exactly n places for each n pulses dialled (Figure 2.5).

Notice that most of the intelligence (the routing of the call, the generating of the ring, the detection of far end busy, etc.) resides with the central exchange. Exchanges control the progress of the call and do not permit, or even recognise, the possibility of within-call divert requests from the terminal. Notice also that you do not call a person, you call a particular pair of wires connected to a specified place. Also,

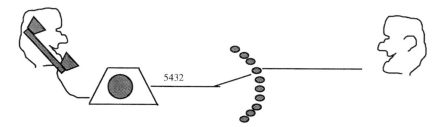

Figure 2.5 Automatic call routing

the connection is completely end-to-end and symmetrical, from a telephone to a telephone, with the expectation that two-way traffic will take place, with similar characteristics in both directions. Finally, during the call, the network is as transparent as possible, providing a fixed width (in terms of signal bandwidth), with minimum end-to-end delay. This is achieved by the operator, or the automatic equipment, establishing a guaranteed end-to-end route.

EVOLUTION OF TELEPHONY ARCHITECTURE – 1960s TO PRESENT DAY

The situation described above lasted essentially into the 1960s, by which time the growth in call volume, the increasing need for national and international traffic and the possibilities arising from improved technology, all led to some significant architectural changes. To take one example, imagine two people, one in Edinburgh and one in Bristol, wish to call someone in London. Under live operator control, they clearly would both ask for a connection to 'London'; that is, they would both use the same name, and one that was, moreover, independent of their location. Automatic trunk exchanges that were rapidly introduced during the 1960s were intended to give the same service (Figure 2.6).

Irrespective of where you were, you dialled the same code for London, 01, followed by a major London exchange number, 543, and then the local number. Obviously, the number dialled cannot correspond one-to-one with the setting up of the switches that control the path through the network. Instead, the dialled digits are captured in a buffer and translated into the digits corresponding to the correct path, and these are subsequently used to control the call. An important change had come over call control and routing: terminals no longer had to send a signal which referred to the fixed settings of the network switches; instead they could request a connection on the basis of

NEW NETWORK ARCHITECTURES

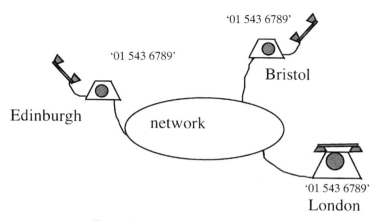

Figure 2.6 Automatic trunk dialling

supplying a number which bore no fixed relation to the switch pattern. This 'call by name' principle, as opposed to 'call by value', to borrow a term in computer terminology, has quite important ramifications. One case is that of 'personal numbering': you can be given a number which refers to you, rather than to the telephone line running into your house. Provided you have some way of indicating to the network that you have moved to another location, then the number dialled can, in principle, be translated to your new location. This is exactly what is done in the case of mobile networks, which maintain continually updated databases of the location of mobile terminals; see the selection on cellular networks for more details. The flexibility of number translation was an early example of software control, and the exchange switching equipment that deployed it was an example of a special-purpose computer. From the beginning of the 1970s there has been an ever-increasing interest in the introduction of computer processing into switching systems, in order to provide a cost-effective and flexible 'intelligent network'. The telecoms companies have had, until perhaps recently, a fairly well-defined and focused view of what they meant by an 'intelligent network'. Why there had recently been a partial loss of nerve we shall discuss later, but let us first look at their original vision.

INTELLIGENT NETWORKS: THE 'TELECOMS VIEW'

The intention was that the intelligent network of the future was still to be predominantly a narrowband, voice network catering for telephony rather than data or multimedia. Because the network was intelligent and built to respond to standardised commands, it would be possible to give a very wide, potentially global set of services offering maximum flexibility to customers (see frame).

The technology to deliver these services is outlined in Figure 2.7. In this design, telephones are still connected to 'local exchanges', some of which will still be 'unintelligent'. However, at some point in the

---

THE INTELLIGENT NETWORK AND THE VIRTUAL BUSINESS

The feasibility of creating 'virtual businesses' – organisations that are combinations of different trading partners, widely geographically separated, that come together rapidly for the duration of a major project, and then go their separate ways – requires a highly flexible infrastructure, particularly in their communications needs. Although located across the world, they require the telephone system that connects all those people working on the same enterprise to give them the functionality they would expect from a single private automatic exchange (PABX) as if they were all located in the same building: they want a corporate directory that allows them short-code dialling to the same number, irrespective of where they are located; they want conferencing to be painless and easy; they want to be able to equip the mobile members of the team with cashless calling facilities; and they want to attribute the call charges to the appropriate functional unit in the business.

If the business involves handling incoming calls from customers, they may want to reroute these calls to a number of enquiry points right across the world, to give 24-hour service and to achieve economies of scale by selectively locating specialist problem solvers. They may require customised recorded announcements, in a range of languages selected on the basis of the caller's location. They may even want advanced, perhaps experimental, services such as speech recognition.

What they will not want is an intensive capital investment and installation programme, particularly if the virtual business is to be flexible, ever-changing and, perhaps, rapidly dissolving. They will prefer to buy in a service package. This is one major market for the Intelligent Network, which can provide all of these services.

NEW NETWORK ARCHITECTURES

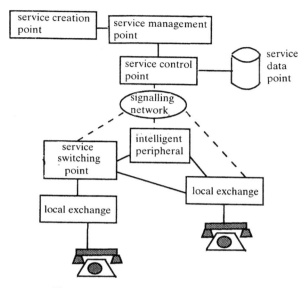

Figure 2.7 The intelligent network

path, the call will reach a 'service switching point' which will set up the further progress of the call (the continuous line) under the command of signalling instructions (the broken lines). What these instructions are is dictated by the software in the 'service control point' and its associated database. The service control point is a high-powered computer that can handle real-time interruptions from the exchanges requesting services. A simple example of its operation is the provision of number translation services: for instance, a business may supply a single number for customer enquiries, but want to handle them at different locations depending on the time of day. The service switching point will have been instructed to signal to the service control point when it receives any such call; the service control point checks the number, realises that number translation is required, retrieves the translation details from the database and returns them to the switching point, which, in turn, connects the call to the appropriate destination exchange. Apart from providing high-speed decision-making, the service control point also performs another important functions: it acts as a buffer between the management and service creation elements and the detailed complexities of switching the calls and of network monitoring and alarms.

The 'application programs' that set up the various services to be performed are written on the computers used for service management and service creation, in high-level scripts that do not require detailed understanding of the switching, etc. For example, we could imagine a programme called 'time of day routing', which simply requested the business customer to complete a table of destination addresses against times of day.

Also shown in the diagram is the 'intelligent peripheral'. Callers and called parties can be connected to such a device, if it were, for example, providing a network answering service. A service could be created that, after normal business hours, switched all calls to the recording machine. In the morning, the business could call into the answering machine and retrieve the calls.

The intelligent network is a major departure from traditional telephony in a number of ways:

- It replaces calls to fixed numbers by calls to flexible 'names'.
- It treats calls in two parts; in particular, a caller can be connected to an intermediate platform such as an intelligent peripheral.
- It allows flexible configuration of the network, to deliver a range of customised services.

Nevertheless, events may have put into doubt some of the earlier certainties of the intelligent network. For instance, it is now realised that networks of the future may have much greater demands for bandwidth and for more complex services. Multimedia terminals are a case in point. The intelligent network will have to be expanded to cope with the flexible switching of circuits with data rates of several megabits, perhaps within an ATM 'packet' network, rather than one consisting of fixed, switched circuits.

But perhaps the biggest impact on the intelligent network may come from the growth in the use of networked computers. We need to look at the parallel developments in the computing industry. We shall find a very different view and a set of attitudes which are in sharp contrast to the opinions held by telecoms professionals (see frame).

THE EVOLUTION OF COMPUTER NETWORKS

Initially, computing networks shared some of the centralised approaches so characteristic of telephone networks. The big computer sat at the centre of the system and access to it was either by physical

> ## YEARS OF NON-COMMUNICATION
>
> Anyone who does not believe that computer and telecommunications operations are still poles apart need only examine the structure of the vast majority of our major companies and observe how uneasy is the relationship between the IT manager and his or her opposite number on the telecommunications side. Frequently they hardly speak. (One senior director remarked, only half-jokingly, that they never spoke because the former would not use the telephone and the latter could not send e-mail.)

media (cards, tape) batch-loaded in a quiet period, or through dumb terminals. Notice, however, that the connection was never symmetrical: it was from a terminal to a different type of device, usually a big machine which may have been serving a number of terminals; also, the terminals could not demand attention whenever they wanted it: the centre polled the terminals according to its own time-plan.

Many of the applications that were developed were not time-critical, at least within a second or so. Payroll and billing systems were early examples and, as performance improved, database query systems, for example on customer help desks, became widespread. Database sizes increased and, for performance and reliability reasons, it became necessary for these large machines to communicate with remote databases over private wires. Programmers and other users of these machines discovered that the networks thus set up provided a convenient means for sending electronic messages: e-mail was born.

The emergence of the low-cost workstation in the 1970s caused a major shift in the architecture of computer systems. 'Client–server' working replaced the centralised, big-box model where most of the intelligence and control resided at the centre, with one where the terminals could request centralised service, such as access control, electronic mailbox management, storage and communication services, etc., from the central machine, but themselves possessed significant processing power and the ability to run applications locally. The network that connected the terminals and the servers was 'local', usually to a building or campus. The data flowing between the terminals and between them and any central service was high speed, but 'bursty'. Because the distances were short, the data-rate capacity of the connecting cable was not usually an issue, and there was no

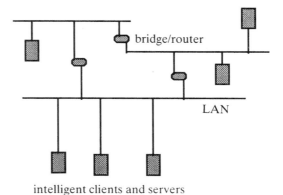

Figure 2.8 Computer networks

need to give each terminal a separate connection. Instead, terminals shared a single physical cable, as part of a local area network or 'LAN'. A number of strategies were developed to cope with the collisions or delays that could occur on the LAN owing to overdemand (see frame).

A need gradually arose to connect more and more networks, over greater and greater distances. The method adopted was to use 'bridges' or 'routers'. These acted as filters to allow signals between the various subnetworks to pass across (Figure 2.8).

The routers are also used to control the passage of traffic across the network, because they are either programmed with network details or learn them by communicating with each other regarding their links to their nearest neighbours.

Routers perform a function analogous to the switches in telephone

---

LOCAL AREA NETWORK PROTOCOLS

The need for the ability to connect up a number of computers within a building or an industrial or academic campus has been noticeable for at least two decades. The basic requirement is that terminals should be easily added and removed, the transport medium (the cable) should be cheap but capable of carrying high rates of traffic, and the system should be manageable despite frequent changes and growth in the number of terminals attached.

There is also a growing need for the network to connect, without problems, to the wider area, in many cases globally. A number of 'standard' ways have been developed, but the most fundamental distinction is between 'non-contending' (usually 'token ring') and 'contending' exemplified by the Ethernet standard.

In token ring, the devices (computers, printers, disks, etc.) are all arranged so that signals on the ring pass through them in turn. There can be a single ring, passing data in one direction, or, as shown, a double ring, that provides terminals with better access and network protection. Bytes of data circulate around the ring. If no terminal is sending data, each of the bytes is a 'token', a standard bit pattern that can be recognised by the terminals. If a device wants to send a message, it grabs one token, modifies it with the data and the name of the receive terminal and replaces it on the ring. This circulates around the ring until the receive terminal detects it. The receiver can then reset the token or flag it as 'refused', perhaps because it is busy or the data is incorrect.

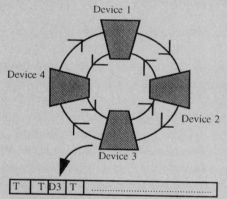

Unlike token ring, Ethernet makes no attempt to arrange things so that devices do not clash for access to the network, which in this case is a simple bus. Devices simultaneously transmit and receive data, and do so at random. Because they can monitor their own signals, as they happen, they can tell whether they have been corrupted by other transmissions. If this occurs, the transmitting terminal waits for a short, random period of time and then retransmits the message. The electrical signal levels have been designed so that it is easy for terminals to determine whether a 'collision' of two or more messages has occurred. If so, the data is ignored and, after a random waiting period, retransmitted.

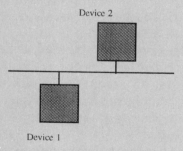

exchanges. However, there is a very distinct difference in the overall philosophy: computer networks have been built from the LAN outwards into wide area networks; the routers (the 'switches') are essentially in the private, local domain of each subnetwork rather than in the public, wide area and they negotiate in a distributed, 'democratic' way with the other routers. In fact, they tend to treat the network as a 'dumb' resource. This makes them very good at finding ways across it, even when there have been path failures, but on the other hand it means that there is very little opportunity to provide a fully defined quality of service. The major deficiency in this respect is in the possibility of long end-to-end delay that cannot even be specified from one burst of data to the next.

WIDE AREA COMPUTER NETWORKS: THE INTERNET

The most significant development in the networking of computers is almost certainly the Internet. From small beginnings in the 1960s, it has become so globalised and grand in scale that it is not possible to put an accurate figure to the number of nodes connected together, or to the total traffic carried. Its phenomenal success, and its significant failings, are a result of the decentralised approach to network routing and 'best-efforts' behaviour rather than defined quality of service. The Internet is an aggregation of the technologies describes above, together with a set of protocols, TCP/IP, that ensure that blocks of data can arrive at the right place and in the right order, but it is more than that: it is an enabler of a number of highly useful services – e-mail, file transfer and interaction with information services – by means of standardised 'browsers', in the concept known as the 'World Wide Web'. As it stands today, however, the major deficiency of the Internet is that it is not possible to give any specification for quality of service, sufficient to meet the demands of services such as real-time multimedia.

In order to understand why, it is necessary to realise the origins of the Internet: it arose through the convergence of two distinctly different strands of thought regarding computer networks. One was the US Department of Defense initiative to develop solutions that would provide communications survivability in times of nuclear attack. The other was the commercial market which was seeking profitable solutions to the connection between local networks of a very heterogeneous nature.

NEW NETWORK ARCHITECTURES

SURVIVABILITY

The US military were concerned that conventional telephony networks, which involved hierarchical connection via centralised telephone exchanges, were vulnerable to attack. It was therefore proposed that the data to be transmitted be broken up into small packets ('datagrams'), each packet labelled with its serial number, source and destination, and the packets fired off independently into the network. At each node in the network, the routing to pass each packet onwards to the next node would be calculated locally, on the basis of knowledge of what was happening near the node (for example, which route out of it was least busy). Packets would eventually get to their intended destination, provided a route existed at all, although they might arrive in any order and experience a significant amount of delay. It is important to realise that there is no intention to maintain a record of the 'best way', from end to end across the network; the idea of any semi-permanent best way was felt to be contrary to what would happen when the bombs began to fall; no route would be guaranteed to survive and the packets would need to find their way through a network whose nodes were constantly collapsing and being rebuilt.

MULTIPLE PROTOCOLS

As well as the military imperative to create survivable networks, there were a number of other market realities: computer networks grew up in a spirit of local service provision and within a fast-moving, competitive, non-standardised market. Consequently, it was inevitable that there would be a variety of communication protocols and little attempt to provide guaranteed quality of service outside the local domain. Nor does the data structure conform to identical layouts. There are a large number of ways that networks can vary, including:

- addressing
- maximum packet size
- error handling
- 'congestion control' (what to do when input to a node exceeds its capacity)
- acknowledgement of receipt of data

and many others. The problem of getting these various types of networks to work together is called 'internetworking', hence the term 'Internet'.

INTERNET PRINCIPLES

To meet this survivable requirement within the commercial realities outlined above, Internet communication involves breaking the data up into datagrams that are of variable length up to 64 Kbytes (although they are usually about 1000–2000 bytes, in practice). These packets of data are frequently broken up by the intervening networks into smaller 'fragments', to fit their maximum size restrictions. These fragments are not recombined until they reach their destination, and one of the complex tasks of the Internet protocols is in adding additional addressing and ordering information onto these fragments, so they can be delivered and reordered successfully. Every host ('computer') and every router on the Internet has a unique address, called its 'IP (Internet protocol) address'. This is a 32-bit address, which is attached to the datagrams to indicate the source and destination of the packet. The address consists of two parts: one part identifies the network to which the host belongs and the other part is a unique number within that network. Networks vary in size and complexity; some are 'private', e.g. a university campus network, while others are provided by regional or national authorities. All networks obtain their IP address from a centralised body, the Network Information Centre. Networks and network addresses can also be split into 'subnetworks', which help with system administration and allows the number range to be extended, rather in the manner that telephone numbers contain a local code, a regional code and a national code.

Suppose a packet arrives at a router. Routers hold tables containing information on some (but not all) of the other networks on the Internet and also some of the addresses to subnets and hosts on its own network. The router compares the packet's destination address with the table and may be able to send it directly to the local LAN where the destination host resides. If the packet is for a remote network, then the router will send it on to the next router recommended in the table. Note that the router does not necessarily send the packet by the most direct route; it may simply send it to a default router which has a bigger set of tables.

Of course, the computers at either end of this route do not wish to know any of the details as to how the packets wend their way over the network. All that they want is reliable transport to be established between them. In order to do this, the two computers establish a 'connection', as shown in Figure 2.9.

The computer that will receive the data is first assumed to be

NEW NETWORK ARCHITECTURES

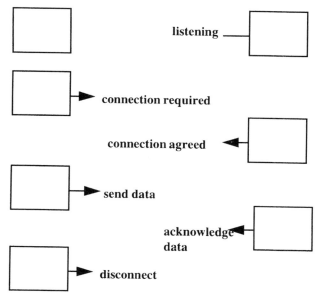

Figure 2.9 An Internet 'connection'

listening in to the network for requests to set up a connection. One is sent from the other computer and it acknowledges it. Both computers are now setting up transmit and receive buffers, counters/timers and error checking. Then one computer transmits some data, hopefully the other host receives it and eventually transmits an 'acknowledge message'. Provided this comes within a specified period of time, the sender assumes that everything is OK and will transmit any further data in the same way.

All this activity is said to take place during a 'connection'. While the connection is in place, the two hosts continually check the validity of the data, sending acknowledgements and retransmissions if there are problems. Interesting things can go wrong: for instance, data packets that have got lost within storage buffers used by a subnet may suddenly pop up after the sender has given up on them and retransmitted the data. Also, during the connection the performance of the route can be assessed and the flow of data adjusted in order to move it through at the maximum possible speed without overload. All of this, and a great deal more, is carried out by the Internet Transmission Control Protocol (TCP). With this protocol in place, it is true to say that the Internet can provide a number of service guarantees, but notably absent is any quality-of-service specification for end-to-end delay. It also turns out that there are a number of scalability issues regarding number ranges and the management of addressing. Some of these will be addressed in next-generation protocols, or are being circumvented by a series of pragmatic fixes. One of these includes the ability to request routers to provide reserved bandwidth paths, provided the routers have the facility and are configured to allow it. This means, in principle, that end-to-end buffering can be minimised and real-time multimedia can be made possible between a receiver and a remote platform. However, the system is not flexible and there can be contention problems if there is significantly more overall demand at the routers than they can cope with. Quality of service, in terms of bandwidth and end-to-end delay, still remains an elusive goal for the Internet. (For example of just how bad this performance can be, see the frame, 'A network performance test'.)

TELECOMS AND COMPUTER NETWORKS – A COMPARISON

It is worth recalling the principal features of computer and telecoms networks we have discussed in the previous sections (Table 2.2).

A NETWORK PERFORMANCE TEST

A number of tests were carried out between a server in southern Italy and user machines in Norway, England and Ireland. From each of the users, 100 short packets of data were sent to the server and a number of statistics gathered. In order to achieve a high level of throughput, without requiring the resending of data, we would expect the system to have a low percentage of data loss and a low and consistent round-trip time. What we found was:

- *Norway to Italy*: 23% of packets were lost or did not arrive within 20 s; of those that did arrive, fastest packets took 0.3 s, slowest 4.0 s, average 1.6 s
- *England to Italy*: 46% lost; fastest 0.4 s, slowest 2 s, average 0.6 s
- *Ireland to Italy*: 13% lost; fastest 0.4 s, slowest 3.4 s, average 1.8 s.

Granted that the path through Italy was known to be poor and involved routing via five routers, we also found significant packet losses and variation when testing between the other sites, where transmission is considered, by Internet standards, to be very good. A telecoms company would find it incredible that this unreliable and highly variable level of service was acceptable.

NEW NETWORK ARCHITECTURES

Table 2.2 Differences between telecom and computer networks

Feature	Telecom network	Computer network
Bandwidth	fixed, usually low	bursty, with high capacity
Multipoint capability	poor	good
Remote access	poor (but improving)	good
Location of intelligence	in the centre	in the terminal
Reliability	very high	poor
Standardisation	good	poor

Bearing in mind that one of the major requirements for future networks will be to provide a fully satisfactory solution for multimedia communication, it turns out that neither of them is quite adequate. (But one possible early defector to 'computer networks' will be facsimile traffic: see the frame below, 'Facsimile, the forgotten service'.)

The biggest problem with the computer network version is that of guaranteeing a low end-to-end delay. The 'collide and try again' approach of Ethernet local area networks is clearly not the correct approach, and it is interesting that the modern approach to building wiring is to eschew the single coaxial cable connecting all the terminals together in favour of putting each terminal on a separate pair of wires taken to a central switch (Figure 2.10).

Essentially, such a system sets up a 'circuit' for each terminal, with guaranteed availability for the terminal, whenever it needs to send a piece of data. To the telecommunications engineer, this is not a novelty; traditional telephone calls have always been offered dedicated channels,

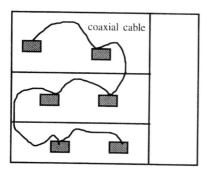
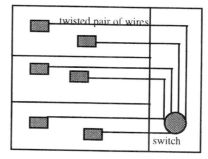

Figure 2.10 Old and new methods of LAN wiring

of given bandwidth or bit-rate. In the early days each channel corresponded to a distinct pair of wires, although in more recent times the channels were combined onto the same physical medium by means of multiplexing techniques such as FDM, TDM and CDMA (see frames pages 100, 33 and 133, 'TDM, FDM,' CDMA). However, each of these channels had a specification which made them less than ideal for voice and almost unacceptable for today's data requirements. The problem with the specification was *bandwidth*, or, to be more accurate, the cost of it. The minimum acceptable bandwidth for speech is around 3 kHz, corresponding to a bit-rate of 10–60 Kbit/s (depending on how the speech is coded). It is reasonably accurate to say that this bandwidth is required most of the time the circuit is active.

Now, the market for voice telephony appears to be acceptable at the current costs of a few pence or cents for a minute or so of long-distance calling. However, a typical local area network operates at bursts of 10 Mbit/s or even more. This bandwidth is 1000 times greater than for telephony and it is unlikely that a market could be sustained for prices in this range, even allowing for bulk discounts! Remember, however, that we said that computer traffic was bursty and required this level of bandwidth only intermittently. If band-

FACSIMILE, THE FORGOTTEN SERVICE

Often forgotten by telecoms and computer experts alike is the great success and revenue-earning potential of facsimile traffic: approximately 20% of all international traffic is fax. Most of this is still generated and received by conventional fax machines and transmitted over the international telephone network. However, in recent years, there has been a progressive shift towards computer-generated fax, from personal computers equipped with fax modem cards. Most of this traffic still passes all the way over telephone circuits, but there is no technical reason why it should do – the low delay and continuously available bandwidth so necessary for speech is not a requirement for fax, which can quite satisfactorily be catered for by the much poorer quality of service, but much cheaper, connections available over the Internat. Internet service providers are beginning to offer dial-in facilities for standard fax machines, offering to deliver the message at much lower cost. This may come to be a bypass threat to a significant fraction of telecoms traffic.

width is going to remain relatively expensive (we shall look at the alternative scenario later), then the solution is obvious: we require to be able to create circuits whose bandwidth can change dynamically, in tune with the demands of the data.

CIRCUIT-SWITCHED AND PACKET-SWITCHED, CONNECTION-BASED AND CONNECTIONLESS

We can see, therefore, that there are at least two major architectural differences between computer and traditional telecom networks: the former throw variable rates and lengths of data 'packets' into a largely uncontrolled wide area network and let the routers work out, on the fly, how to get them to the destination. On the other hand, telecom networks carefully work out the entire route beforehand and then set up a fixed-bandwidth end-to-end circuit. Adding TCP to the basic IP computer networks creates a 'connection', between terminals, that avoids the problems of disordered or lost packets, and thus some level of quality control can be achieved; but, because there is no attempt to create an end-to-end circuit, there is no way that lengthy and variable delays can be avoided. As shown in Figure 2.11, telephony circuits consist of setting up a continuous track (a 'switched circuit') from start to finish, before the train is allowed to proceed, whereas traditional computer networks just throw packets of data 'over the wall', from router to router, with most intermediate stages ignorant of the best way to the final destination. The train has better quality of service, but its capacity is fixed.

Logically, then, we might see the beginning of a useful convergence: what about a system which establishes an end-to-end connection, but is capable of handling variable bit-rate and variable length data packets?

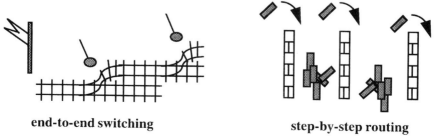

end-to-end switching   **step-by-step routing**

Figure 2.11 Telephone 'railway' and Internet 'wall-game'

ATM – TELECOMS' ANSWER TO THE VARIABLE BANDWIDTH PROBLEM

The dominant architecture of telecommunications networks is, to this day, based on the concept of the fixed-bandwidth, semi-permanently switched circuit. The simplest example is the telephone call, where the local connection between the customer and the first switching point (e.g. a local exchange) has a fixed bandwidth (approximately 300–3000 Hz) for an analogue call, or 64 Kbit/s for a digital, ISDN call). Throughout the network path that continues on to the called party, the switching and the transmission equipment are all designed to preserve this constraint, which then impacts on the cost and price structure to the operator and customer, respectively. Business customers may hire wider-band circuits, to connect their PABX's or computers, but these two are constrained to a fixed bit-rate within well-defined digital hierarchies. Many of these circuits are permanently 'nailed up' from origin to destination by infrequently changed data in exchange software, or even by real jumper-wires. The price is calculated as a fairly simple function of

- fixed bit-rate
- distance
- duration of contract.

The true relationship between distance and cost to the network operator is becoming progressively much weaker and prices will eventually flatten accordingly, but costs for bit-rate and duration are much less flexible under the constraints of the current architectures. This is bad news for the bursty, intermittent signals from computer terminals. Network operators believe that the medium term (which could take us as far as 2020, in their planning terms) lies with Asynchronous Transfer Mode (ATM), which provides a way to set up and take down, very rapidly, circuits with variable bandwidth but also possessing low end-to-end delay.

ATM works on the principle of statistical multiplexing: the fact that combining a number of individually fluctuating channels results in the creation of a larger channel with lower percentage fluctuation which can thus be more heavily loaded without overloading (see frame). Suppose we have a number of individual channels, $A, b, \ldots, z$, which will not, in general, contain data at a steady rate. We combine them through a 'multiplexor' onto one transmission circuit, on the basis of allocating each channel a fixed-length chunk of data (called a

Figure 2.12 ATM multiplexing at one instant

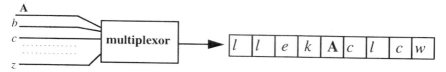

Figure 2.13 ATM multiplexing at another instant

'cell'), depending on need. So, if we monitored the output of the multiplexor, we might see at one instant (Figure 2.12) that source A is relatively busy. At a different time, which could be milliseconds or hours away, we might see the picture shown in Figure 2.13. This time, A has required only one-fifth of the previous data rate (and might expect to pay accordingly).

STATISTICAL MULTIPLEXING AND THE LAW OF LARGE NUMBERS

Consider a source of data whose rate varies with time. On average the data-rate is m and we also have some measure of the variation, s. (This could, for instance, be the 'standard deviation').

If we want to transmit the signal over a fixed bandwidth channel, we need to make allowances, not just for the mean rate, but also for the variation. This means we need a channel bit-rate which is proportional to m and to s.

> Suppose we have another signal, which, for simplicity, also has mean *m* and variation *s* and whose variations are independent of the variations of the first signal.
>
> At some times, the two signals will both be running above average rate, at which times their variations will add together directly. But at other times, one will be high and the other low.
>
> Variations do not add like means. In fact, under some circumstances and particularly when *N* is large, we can say  that, while *N* signals, each of average rate *m*, has a total average rate of $N \times m$, the same combination has a variation that is only $N^{1/2} \times v$. Thus 100 independent channels requiring 1 Mbit/s for 0.7 Mbit/s of average rate plus 0.3 Mbit/s for variation require 100 Mbit/s, whereas a multiplexed version requires only 100×0.7 Mbit/s + 10 (= $100^{1/2}$) × 0.3 Mbit/s = 73 Mbit/s.

By breaking the data into cells, the system can therefore cope economically with variations in data rate. Selection of the cell size – the number of bits per cell – is a trade-off between a number of factors: the longer the cell, the greater delay that occurs in putting the cell together and the processing of it through a multiplexor; the smaller the cell, the greater the ratio of control information to 'real' data.

ATM uses a fixed cell size of 53 bytes, of which 5 bytes are control data and the remaining 48 are message data. (Strictly, 'message data' can also cover higher-level control data, for example for fitting data from computer systems using the variable-length data packets characteristic of the Internet TCP/IP format, but at the call level only 5 bytes are used for routing it through the network.)

Apart from its variable bit-rate and the statistical nature of its operation (the latter we discuss further below), ATM is otherwise very like a conventional telecommunications architecture (Figure 2.14).

A connection between A and B is made across a series of ATM switches within a digital hierarchy of speeds (only some possible examples are shown). Each connection is described in the 5-byte control data in terms of a 'path' – that is, a route between two switches – and, since each path will carry a large number of separate circuits, by a 'channel' description within that path. ATM switches maintain tables of which channels on which links are linked together and they rewrite this information into the control data as the cells are switched

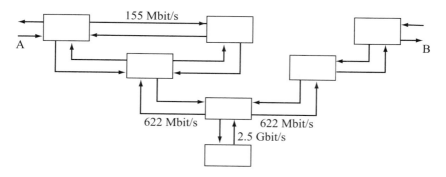

Figure 2.14 ATM switch hierarchy

through the network. The path is worked out, end-to-end, before cells of data are accepted.

PROBLEMS WITH ATM

ATM is a compromise and is likely to be a successful one, in the medium term, but it is not without its problems. One strong argument for designing ATM in this manner was that it lent itself easily to high-speed, hardware implementation rather better than alternatives, but this argument has been somewhat overtaken by the pace of technology. There are problems with fitting in some of the computer protocols currently in use. ATM is not easy to adapt for use in a multipoint to multipoint transmission, for example in the case of a distributed team that wanted to hold a video-conference, viewing each other from a number of sites. There are also some problems of cost and efficiency for low-speed, commoditised ATM delivered direct to the terminal.

These problems and several others can be worked round, but there is one much bigger problem: it is very difficult to define quality of service in any network based on statistical multiplexing.

We have said that statistical multiplexors can take advantage of the fact that, on average, the percentage variation in the demand for bandwidth is less across a number of circuits than it is for them individually and we can, therefore, run our large transmission systems closer to maximum capacity, without danger of overload. The problem is that this is only *on average* correct. On some occasions, too many of the individual circuits will all be generating data well above their long-term average, and things are going to overflow. The statistics of the traffic cause enormous problems directly to the network provider, in terms of getting the transmission quality to an acceptable standard at an economic cost, and indirectly to the customers, who need know nothing about the solutions adopted but are critically interested in the price. Consider, for example, the 'simple' case of customers who agree that they will never send data at rates higher than 100 Mbit/s. But do they mean one single 100 Mbit packet every second, or 1 Kbit every 10 microseconds? Clearly, the effects on the network will be different.

ATM FLOW CONTROL AND SERVICE POLICIES

A tremendous effort has been and is going on into developing ways of minimising the consequences of these statistical fluctuations. An obvious solution is to control the flow of data entering the network. One widely investigated technique is to put additional storage at the input of the ATM switches and use it to buffer data into a more regular stream. Although called the 'leaky bucket' principle, it is more akin to a funnel that can accept a flow however irregular, but equipped with a tap which opens at a regular rate R, to release a fixed 'drip' of data (Figure 2.15).

It is possible to leave the management of this buffering to the customer, rather than build it into the network. The network provider can easily police the behaviour of the customer by periodically monitoring the value of R: if this becomes faster than the purchased data-rate, the customer is violating the agreement and, for example, can be switched to a punitive tariff regime.

The ATM network provider's task then becomes that of supplying a pipe which is big enough to take the inputs from a number of funnels but small enough to be efficiently filled, as shown in Figure 2.16.

Unfortunately, all this is beginning to give rise to a great deal of uncertainty and difficult decision-making for customer and network provider alike. We begin to define a range of services, with a range of penalties for over- or under-specifying our requirements (Table 2.3).

The table shows only a few of the possible scenarios, but they make it clear that an ATM service will have many more options than traditional, fixed circuit telephony. For instance, cases 2 and 3 in the table both give the provider a number of choices: charge more or lose traffic, adjust tariffs for time of day or other slack periods. Whilst this might be seen by lovers of unrestrained competition as a good thing,

NEW NETWORK ARCHITECTURES

DIAGS 7

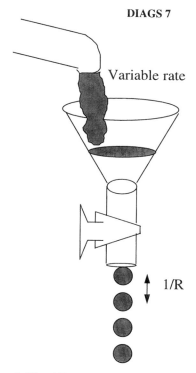

Figure 2.15 ATM 'leaky bucket' flow control

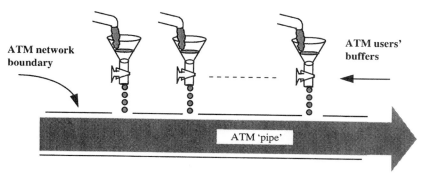

Figure 2.16 ATM network access flow control

we also have to be aware of some danger signals: although a simple stock-market approach to buying bandwidth might benefit customers in the short term, over routes served by only one carrier, conversely a network provider who gets its tariffing marginally wrong could lose the majority of its custom overnight. What happens in the case of

Table 2.3 Some ATM service options

Customer require	Network provider's response
Case 1: need guaranteed bandwidth and end-to-end fixed and short delay e.g. for voice), with set limits on data loss.	I shall charge you my top rate and it will not be inexpensive.
Case 2: I estimate my bandwidth to be less than a certain value, but I need the other conditions.	You can have it cheaper, but cheat on me and it will be very expensive, or I may choose to lose your data.
Case 3: I'll take what you've got.	Bargain-basement prices, but performance will vary depending on time of day and my success in attracting premium customers.

traffic across a number of differently tariffed networks? ATM has the potential to create a greater degree of market uncertainty for wideband services. This will probably have an inhibiting effect on its introduction. However, there is little doubt that statistical systems like ATM will form the backbone of a large percentage of our networks over the next two decades or so.

PROBLEMS WITH COMPUTER NETWORK PROTOCOLS

Although the computer-network community is not particularly enamoured of ATM, for some of the reasons given above, it is true to say that ATM may win by default, for many applications, because there is no immediately obvious alternative. We have already mentioned TCP/IP, the widely used protocol for intercomputer networking. It is widely recognised by the community that this protocol is (a) inadequate at present for a number of tasks, yet (b) being used by an exponentially rising number of users. Even though the telecommunications community has an equal dislike of TCP/IP as the computer camp has for ATM, it does seem that some form of TCP/IP will be a dominant force in networks over the next few years, even though it has a number of problems. One of those is TCP/IP's ability to perform at speed. Computer protocols were originally designed in the days when networks ran at sub-megabit per second rates. In the chapter which follows, we are going to discuss the possibility of wide area transmission at gigabits per second. Not unnaturally, this is going to pose a problem for existing practices. A few pages back, we described how data packages were acknowledged by the receiving computer and, provided this acknowledgement got back to the sender within a certain time, then it did not need to resend data. In fact, this time limit was determined by the size of the sending computer's output buffer. If the buffer emptied before the acknowledgement was returned, then the sender had to take appropriate action. Suppose the buffer is N bits long and the transmission rate is R. The buffer will empty in time N/R. This time must be at least as long as the time taken for the first bit of data to go from sender to receiver and for the receiver to acknowledge it. Suppose we have a 500 km line. Even if this is a single optical fibre, with no intervening switches or routers, the round-trip time will be of the order of 1/200 s. So:

$$N/R \text{ must be greater than } 1/200$$

that is, N must be greater than $R/200$. This is no problem if the data-rate is, say, 2 Mbit/s. We can easily have a buffer of 10 kilobits. If the data-rate goes up to gigabit per second rates, then we begin to talk about quite large buffer sizes. There is another problem. Suppose we have an error in the data. This will not be relayed back to the sender until an enormous amount of data has been received. Currently, protocols tend to retransmit multiple packets after an error is detected. This seriously limits transmission speeds, when we are delay-limited, rather than bit-rate-limited, by the transmission system.

There are other protocol issues that require changing, if computer systems are to make full use of ultra-high-speed networks. For instance, many of the current methods are heavily biased towards recovery after relatively frequent transmission errors. New transmission systems based on fibre will greatly reduce the error rates encountered and this should be capitalised on. In general, computer protocol engineering needs to look beyond the relatively short-term gains to be achieved by modifications to TCP/IP and so on.

PROTOCOL CONVERGENCE?

Where therefore do we go from here in trying to reconcile the protocols that were originally designed to meet different requirements at the time when bit rates were more modest and the requirement for real-time multimedia was not considered significant? One possibility is out-and-out war between the end-to-end switchers and the hop-by-hop routers. This is a real possibility, albeit a rather frightening one, but it is perhaps better to concentrate on more hopeful outcomes. It is possible that some accommodation can be achieved.

In particular, we should remember that protocols achieve their greatest elaboration and least flexibility when they are trying to squeeze a quart into a pint pot. IP, the very simple datagram service, had to be enhanced by TCP, in order to achieve some level of quality of service, in the hostile, error-prone, bit-rate-limited, unregulated Internet. ATM, on the other hand, has a rather rigid structure, partly at least because of speed restrictions on the switching fabric. Optical fibre transmission, with its enormous bandwidth and very low error rates, may allow us to relax the strait-jacketed protocols. One possibility that is emerging is for IP packets to be run on top of an ATM-based network, with the routing being done using IP methods but ATM principles being used to control the flow of data into the network, using the leaky bucket principle. This is a bit of a 'bodge', but is, at least, a sensible compromise for some types of traffic, and could be the seeds of a new set of protocols. Some research teams have suggested that it is possible and desirable to integrate the data routing of IP and ATM into one scheme. They also suggest that this combined scheme should, however, offer two classes of service: one which does a best-efforts delivery and the other which promises to meet specified quality of service.

THE SPECIAL CASE OF THE LOCAL LOOP

The majority of the transmission cost in a telecommunications network is not in the long-haul routes, but in the cables to our homes and offices. Despite the radical changes in the long-haul – coaxial cable replacing pairs of wires and, more recently, the introduction of fibre optics and satellite – the vast majority of telephone connections are carried 'the last mile' in the local loop over copper pairs of wires. This is even true in the case of telephony delivered by CATV companies, who use coaxial cable for delivery of TV into the home, but run the telephony connection separately, on copper pairs. Only in the case of multiline business premises are optical fibre or radio links becoming the predominant technology. It is currently considered too expensive to rip out existing domestic cable and replace it with fibre, and the optical transmitters and receivers are still too expensive on a single-line basis, if one is providing only telephony.

Coaxial Cable Solutions

CATV companies have some advantages, therefore, over telecommunications suppliers, in that they usually provide coaxial cable connections direct into the home. Originally these were simple one-way transmissions of multiple channels of TV identically into every home (Figure 2.17). This is, essentially, 'TV broadcasting in a pipe', all the TV channels being broadcast from a central hub onto a master cable that was broken out into smaller cables, with one-way amplifiers blocking off the possibility of return messages.

However, the situation is rapidly changing. A street of houses is served by an optical fibre which carries all the programming channels. A backward path is also provided on this fibre, or on a separate one. The fibre is terminated in a street cabinet, which holds electro-optical equipment that splits out the signals destined for each individual house. This individual traffic is carried into each house on a separate 'tail' of flexible coaxial cable. As well as having enough bandwidth to carry in the TV channels, the coaxial cable also can handle, typically, 2 Mbit/s data to and from a 'cable modem'. This allows the householder to have very fast Internet access, over the modems which can cost less than £100.

THE TELECOMS ANSWER: 'VIDEO-ON-DEMAND ADSL' OVER COPPER

Naturally, telecommunications companies have been concerned to counter the threat from cable companies who are not just selling

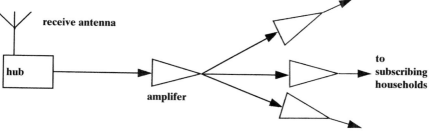

Figure 2.17 Early cable TV scheme

access to CATV, but also frequently packaging it with telephony. One solution, 'video-on-demand using asymmetric digital subscriber loop (ADSL)' is a system for sending TV-quality images over the existing copper pairs of wires that carry domestic telephony.

There are two problems with the copper wire used for carrying local loop traffic: firstly, it is limited in the bandwidth/bit-rate it can carry. This is a function of distance: a signal at 1000 bit/s is hardly attenuated at all at a distance of 1 km, but a 2 Mbit/s signal will be reduced to one hundredth of its size, or even less, after travelling the same distance. This in itself would not be a problem; it is easy enough to design amplifiers which can restore such signals to their original level, but this is where the second problem with copper pairs comes in: induced noise. Copper pairs are rather good at picking up noise from other sources, particularly those on other pairs in a typical multipair cable. This problem can be reduced somewhat by twisting the pair of wires together, but it cannot be entirely eliminated and, unfortunately, it gets worse as the bit-rate increases. Our 1000 bit/s signal will induce a version of itself into an adjacent pair of wires, that is only a few thousandths of itself, whereas a 2 Mbit/s signal is transferred across as several parts in 100, Referring to Figure 2.18, receiver B receives a much attenuated 2 Mbit/s signal from transmitter A. Close to B is transmitter C, whose signal is coupled into the pair of wires connected to B. Because C is close to B, its signal (called 'near-end crosstalk') will not be attenuated much and therefore completely drowns out the faint signal from A. But consider transmitter D. It too is generating a 'crosstalk' signal into the pair connecting A and B. However, it is as far away from B as is A, and its signal

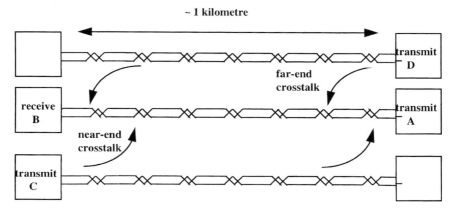

Figure 2.18 Near-end and far-end crosstalk

will suffer the same attenuation as the signal from A. So, the signal received at B will always consist mainly of A's signal and little of D's.

Video-on-demand ADSL (at least in its original design) relies on the fact that high bit-rate traffic is always one-way – from the service provider to the domestic customer – unlike the symmetrical case of telephony, and therefore there are no cases of near-end crosstalk. (There is no interference from the two-way telephony signal, because the video signal operates within a frequency band above that of the speech. Similarly, the narrowband reverse direction channel, provided to allow customers to send control messages, is also out of band with both.) Typically, two or three bit streams of 2 Mbit/s, each corresponding to an individual channel of TV quality, can be sent over a distance of several kilometres, covering the vast majority of connection links between local telephone exchanges and the customers they serve.

Originally, this approach was seen as a way of offering competing services to cable distribution of movies on demand: customers individually accessed a computer server equipped with massive disk storage containing a selection of movies and TV recordings. Each customer's home terminal, a "set-top box' associated with a standard TV set, created a specific read operation to the disk drive, requesting programme content to be down loaded continuously. There were legal advantages to this individual 'on-demand' access (it was not a 'broadcast') and it meant that programmes could start and pause as the individual customer wanted. Home shopping is another application: interaction between the set-top box and the server would allow customers to browse through video shopping catalogues.

However, an alternative use for ADSL has emerged: high-speed access to the Internet. In exactly the way that cable modems are proposed as a way of achieving megabit per second performance by the cable companies, the telecommunications companies are considering ADSL. They are also looking at a variant of the cable company hybrid fibre–coax solution – hybrid fibre–copper – where there is a street feed over fibre, with the existing copper pair completing the final delivery to the premises. Standards for bit rates of up to 51 Mbit/s are currently being defined.

INTELLIGENT TERMINALS

We said that the intelligent network concept has been shaken somewhat by the growth in importance of networked computing. Perhaps the major impact will be from the existence of intelligence in the terminal. Even in the intelligent network, telephones are expected to be relatively 'dumb'. It will be necessary for the telephone to interact with the service switching point, for example, to retrieve messages from an intelligent peripheral acting as a networked answering machine, but these interactions are assumed to be relatively uncomplicated and could be evoked by using standard multi-frequency tones generated from a simple telephone's keypad.

The tones are interpreted at the service switching point and converted into messages carried over the signalling channel, to the service control point. The telephone interacts with only one network; its signalling (tones) and traffic (speech) are, for at least part of the route, carried over the same path, and the intelligent processing is done by the central intelligence of the network.

But in the computer model of a network, the terminals are intelligent and the network is dumb. What happens when telephony is exposed to this model? Take the example shown in Figure 2.19. Alice has her telephone connected via her computer. The computer is connected to the Internet and also to the local telephone company's line. She dials Bob's number. Her computer has been in contact with an Internet service provider who supplies a list of the cheapest telecoms tariff for the long-distance route together with the access code

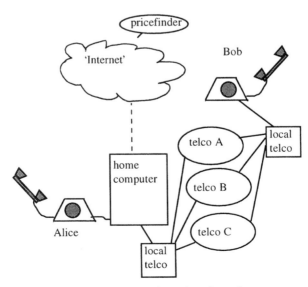

Figure 2.19 Least-cost call routing

that has to be dialled. Alice's computer does a number translation from the number she dialled to that of the cheapest route and then dials that number. What Alice has done is to bypass the central control of the service creation environment, the service control point and, in fact, the entire architecture of the telecoms company's intelligent network. (At the time this was first written, this scenario was considered to be only a theoretical possibility and rather discounted by traditionalists. It is now a reality.)

In the ultimate, the terminal could post a request for tender each time it wanted to make a call and could then sit back waiting for the best offer.

LOW-COST VIRTUAL PRIVATE NETWORKS

There are other services that can bypass some of the functions once expected to be the preserve of a centralised intelligent network. One further example is as follows. Imagine setting up a virtual private network, consisting of a telephone and data network, where people's addresses and telephone numbers are logically assigned for the duration of a project. With simple telephones as the only voice terminals this would require configuration of a network switch, the value-added contribution of the network provider, but if telephony is mediated by the workstation, then it can perform number translation and even hold call records for interdivisional charging. In the next few years the power of personal computers will be sufficiently great to allow them to carry out much of the control of the call, if a mechanism exists for mapping this intelligence onto the switching. If this is the case, ownership of the routing largely passes from the carrier to the customer.

How this will happen in detail is up for debate. In the examples given, the computer has carried out intelligent control independently of the telecoms network, in some cases making use of a parallel data network (the Internet). This is an extreme reaction to the earlier centralised model of control. An alternative might be a more co-operative regime where the local intelligence of the terminal interacted with the service control point of the intelligent network. A logical option might be to ask the user's terminal to configure the features and requirements to meet individual needs at individual instances, for instance, to find the cheapest available provider of a video-conference, and use the central intelligence of the network for issues of scale, such as configuration of global number plans for a virtual private network, the centralised distribution of voice announcements, universal access between fixed and mobile services, and so on.

INFORMATION NETWORKS

A further conceptual change required from telecommunications companies is in coming to terms with the fact that communications is not necessarily about people talking to each other or even for computers doing the same. In particular, it is not necessarily about a simple setting up and closing down of a connection, with symmetrical flows of data between the two parties. In many cases, the interaction will be highly unequal, between 'consumers' of information and its 'suppliers' (Figure 2.20).

There will be a large number of consumers, equipped with relatively cheap terminals – home computers, personal (mobile) organisers, multimedia telephones, digital TV sets – who will access a relatively small number of suppliers, typically larger computers that contain multimedia information or entertainment, that can be supplied to individual consumers, 'on demand'. Data flows in such applications are largely one-way: the consumer gives relatively simple, short, slow commands (by mouse or keyboard now, perhaps by voice in the future) to the supplier, who returns vast volumes of data at a very high rate – typically, perhaps 32 Kbit/s of requests (i.e. the capacity of one voice-grade telephone line) against at least 2 Mbit/s from the supplier (for TV-quality video and sound).

A number of changes to the way telecommunications companies

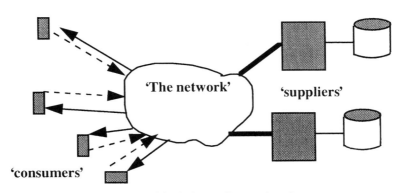

Figure 2.20 Information network

need look at standards arise from this movement away from real-time voice: the needs of entertainment and information services are very different from that of interactive voice and videophone. Consider the last: the video image has to be captured, coded and decoded equally at both ends of the call, and with as little delay as possible. This is not true in entertainment services; because there are fewer suppliers (who encode the material) than receivers (who decode it), it is advantageous to use expensive coding equipment for the former, and inexpensive decoders. Also, there is no pressing need to encode without delay; the exchange is not real-time interactive.

'Information services', by which we mean extensions to the content and services currently available on the Internet, raise another complication: what protocols will we use for delivering the signals to our home computers and set-top boxes? The current favourite is to use Internet protocols. These, as we have seen, however, are still somewhat deficient in terms of quality of service. There are likely to be significant developments in this area over the next decade, before the issue is finally resolved.

MOBILE CODE AND NETWORK 'AGENTS'

Despite the significant difference between current computer and telecommunications architectures, they both tend to localise their methods of control. Telecommunications networks, as we have seen, have guarded very jealously the authority of the centre to possess the intelligence and the power to control things. Computer networks allow more autonomy to the 'peripherals' (the computers at the ends of the network) but the control possessed by the terminals is strictly over their own domains. When computers cooperate on a task, they tend to do so by passing each other data, i.e. 'data' in the sense that it is used as such by the remote processor, not as 'code'. The machine receiving the data is still completely in control of its local processes.

Recently there have been interesting developments in approaches to communication across networks, and also in network management, that have employed a rather different approach to communication and control architectures, one that employs mobile software agents – pieces of executable code – that traverse the network and execute their instructions on its control elements and the computers and other terminals at the ends.

At this point, a warning: there is considerable discussion about precisely what is meant by an 'intelligent agent' – the term is used in many ways and in order to follow fashion, overuse has tended to devalue its meaning. 'Agent' is more of a phenomenalistic and anthropomorphic term than a description of the interior workings; 'agents' are thought of in terms of artificial intelligence entities that in some way behave like human 'agents' – as autonomous individuals that negotiate and communicate and move about. The metaphor is striking and useful, but must not be overdone. Instead, we shall look at some applications.

The 'Singing Telegram'

This very simple example is possible today, using programming languages such as Java (see the frame on 'Portable code'): Alice wants to send a birthday message to Bob, to appear on Bob's computer terminal. She connects to a World Wide Web site on the Internet that specialises in greeting cards. She chooses one that she likes and fills in an online form that will personalise the greeting. When Bob next accesses his terminal, the greeting card will have been sent in the form of a small portion of code and data. When Bob activates the greeting, probably within a Net Browser, the code will be automatically translated on his machine and will be executed: perhaps a dancing bottle will move across the screen, singing.

> 'To the birthday boy within his palace,
> Let me take you for a chalice,
> Of wine at Rizi's,
> Love from Alice'.

(Unfortunately technology does not automatically guarantee taste or talent.) Bob can then click on the 'yes, let's!' button that goes with the greeting, to automatically return an acknowledgement.

Agents for Mobile Users

The above example demonstrated that terminals could operate as pairs of distributed computing engines, with one terminal running code supplied by another. In that example, the network was involved only for transporting the code. Now consider the case of a business executive on the move: she wants to arrange a new venue for a meeting with a client and also plan travel and accommodation to coincide with the meeting. The problem is, she is shortly going to

> **PORTABLE CODE**
>
> In the sections on new access terminals, we discuss the possibility of 'minimalist' computers which do not permanently hold code for all the applications and utilities that are frequently used; instead, they download code that is not just data, but also executable programs as and when they are required. In recent years a number of programming languages have been developed that make this process feasible. An early example, from General Magic, was intended to provide this facility, in particular from handheld mobile units, which firstly would not have a large memory space available for programming and secondly, because of their association with mobile telephones, would be in only intermittent contact with the rest of the network. The idea was to be able to launch executable code into the network, targeted at online servers that were equipped with compatible software, and then retrieve the results at a later time. In practice, the concept of mobile code has caught on more quickly in the large installed base of 'conventional' personal computers, particularly in the case of the Java programming language, originally pioneered by Sun Microsystems. Users can download small portions of code which can be either 'compiled' (translated into a set of machine code instructions will full access to the computer's facilities or 'interpreted' (converted to a subset of the full instruction set and therefore less likely to contain viruses, etc.).

board an aeroplane inside which communication will not be possible. There is enough time to use a personal communicator to fire off an agent into the network, bearing a set of instructions: the agent has to communicate with the client's electronic diary to find a meeting date, then contact a travel office to book flights, hotels, etc., and possibly transfer securely payment from the executive's bank. Once this is done (or, alternatively, when it is discovered that alternative arrangements are required), the executive must be contacted to give confirmation. Much of the processing will again have been carried out at the periphery, but there will be a need to locate the various services (bank, travel agent, etc.) using 'Yellow Pages' directory services and there will need to be quite a complex call redirection service. There is considerable potential for the 'intelligent network' concept to find an application here.

Agents Within the Network

There are applications wholly within the network, too, for intelligent agents: traditional approaches to the management of networks – for instance, trying to balance the amount of traffic that flows along any one transmission path – are becoming the major component of modern network costs. It has been pointed out that, in terms of lines of text, the software code for a small telephone exchange is double the length of the complete works of Shakespeare, the space shuttle has double that again, but a typical network management system requires 12 times as much. Network management is clearly an expensive task, both in its development and in its maintenance. Some experts have predicted that the costs of future networks will be dominated by the software costs, unless we can find an alternative. One possibility is suggested in Figure 2.21.

The graph predicts both an exponential rise in the proportion of expenditure attributed to software and also a possible solution, labelled 'self organisation'. This approach relies upon creating a colony of 'mobile agents' which move from one network node to another, communicating with each other through very simple messages that convey a history of the traffic passing across the links and through the nodes. From this information the nodes can calculate how best to change their routing patterns in order to optimise the flow of traffic. There is no central control system; instead, the nodes behave like cooperative individuals and the behaviour is 'self organising'. So-called 'load agents'

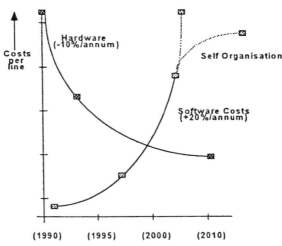

Figure 2.21 The software problem and a solution (source: BT Labs)

are sent to each of these nodes, by 'parent agents' who have been wandering at random around the network, collecting information on how much traffic is arriving at the nodes and the node capacity. When a load agent arrives on a node, it updates the route from that node to all the other nodes by modifying appropriate routing tables throughout the network. Once it has visited a node and made the appropriate modifications, then it moves to another node and repeats the calculation. The agent has acted 'autonomously' and has communicated with other agents only via the changes it has made to the routing table.

On its travels, the parent agent may find 'crashed agents' – agents whose time-stamp to complete the task has exceeded a threshold. It will remove these agents, reset the records and launch another agent. Thus nodes may have several agents working on their routing tables at once. Parents also require to be managed. On some nodes reside static processes which hold a preset figure for the number of parents they will try to maintain in the network. Whenever a parent visits a node, the process reads its name and when it started. If there are many parents on the network, some measure of culling is required. Usually it will be the youngest agents that are removed, since their historical records are least valuable. Sometimes, however, analysis of the visiting patterns of agents revealed that they are 'crashed'. In this case, irrespective of age, they will be removed.

This design is very robust and can handle many process errors, link and node failures and need not be equipped with accurate network statistics at initiation time. The code required to implement it is relatively compact and highly scaleable – all that is required is to replicate more agents.

POSITIVE ASPECTS OF DISTRIBUTED AGENTS

We can see from the examples above that there are a number of very positive reasons for the introduction of agents into distributed networks:

- They are relatively small units of code that communicate with other agents and processes through simple, well-defined messages – thus their operations are individually easier to verify than those of monolithic programs.
- This also makes them suitable for designing applications that run across a variety of platforms from a plurality of service providers.

- Mobile agents can be injected into a network from terminals that are not continuously in reliable contact with the network (e.g. from mobile units).
- The agents are distributed across the network and its terminals; thus, inherently they mimic the network's resilience and redundancy.

MOBILE AGENTS – UNRESOLVED ISSUES

There are, however, a number of question marks still hanging over the wholesale use of mobile agents. Firstly, there is the question of security: a piece of mobile code sent to you by someone else, that runs on your computer, is perilously close to being a computer virus. How do you know that what you have permitted to run on your machine is what you think it is? How do you protect yourself from it doing damage to your files or, perhaps more serious because it is not so obvious, how do you stop it sending details from them, back over the network to the sender?

It is the flexibility of computer designs that allow them to take advantage of the potential for the running of remote code, but this flexibility is also a security weakness. As it stands today, there are risks created by mobile code running on today's machines. Some of these will be removed in time, but mobile code will always provide a relatively equal battleground for the hacker and the security guard.

Secondly, there is the issue of guaranteeing performance: in many cases it is possible to give examples of agent-based design producing solutions that are more cost-effective than traditional methods. But these are on a case-by-case basis. It is extremely difficult to give a reasonably rigorous proof that the agent solution will be 'generally' better, or indeed reliable. The problem is that the more intelligent, cooperative agents that are required for realistically sized problems possess adaptive structures and exhibit 'emergent behaviour' – that is, behaviour whose properties are not easily deducible before they are set loose and self-trained on the real data. There will have to be some very impressive demonstrations of cost-saving and performance before the relatively conservative network operators will carry out sizeable, live trials of agent-based software, on critical tasks in the network management domain.

DISTRIBUTED COMPUTING ARCHITECTURES

The creation of 'virtual organisations' – business enterprises where different trading partners, perhaps on other continents, combine together to create the end-to-end value chains necessary to market a new product or service – will demand a concerted approach to the creation of distributed computing environments to support the enterprise. There must be a degree of cooperation between the applications that run on computers on the different sites (some of which may belong to other companies and employ equipment from different manufacturers), there will be 'resource sharing' of powerful computers, storage or peripherals, and all of this must be as transparent as possible to the user and be capable of fail-safe operation in the event of a link or machine fault.

Quite a large amount of distributed computing goes on at present, of course: computers in different companies communicate automatically with each other in the form of e-mails or electronic data interchange (EDI) transactions (see frame). The interfaces between the

---

EDI – A FRAMEWORK FOR INTERBUSINESS TRADING

Electronic data interchange (EDI) has been a quiet revolution in interbusiness trading, replacing the expensive and unreliable manual handling of orders, invoices, customs and excise dockets, etc., by standardised electronic forms. EDI is not fundamentally a technology (although computers and communication networks are required to fulfil it). Rather, it is a systematic approach to defining the significant operations that make up specific, standard interbusiness tasks – ordering, supplying and paying for goods, for example – defining the data that needs to be processed for these tasks, defining the formats that will be used to represent that data, and finally getting industry agreement to all of that, on a task-domain basis, through the appropriate trade association or intergovernmental body. One of the claims made for the business gains to be made by firms that introduce EDI has nothing to do with technology – it is simply that the very activity of examining business processes to see how they fit into standard EDI models is, in itself, a way of critically assessing the costs and values of the components of the process that can lead to streamlining and efficiency measures.

computers are, however, restricted to the passing of messages, that is, 'data', without the power to control processes, i.e. 'programs', running on the other machine.

There are also large networks of computers that operate between widely separated sites of individual companies, and these can initiate processes on remote machines, but under highly controlled circumstances and only on machines of compatible design.

Some degree of cooperative computing occurs over the Internet; at least two examples are worth noting in this context. Users who access remote 'Web servers' by typing in a URL can, provided the server permits them to do so, send parameters to the server that will start processes running on the server, using a message described in a standard format, the 'Common Gateway Interface' (CGI) protocol. The second example is in the translation of the URLs themselves; URLs are names, not addresses. To be of use in connecting between a client computer requesting the service and a server offering it, the server's URL must be converted to an address (which is then used by the network routers in order to work out the correct network paths to follow). Domain Name Servers are provided in a number of the subnetworks that comprise the Internet to perform this function. Software in the individual client terminals can store some name to address information, but, when a new URL is used, the client contacts the Domain Name Server for the translation. We shall see later how the principles of translation services such as this play an important part in distributed computing design.

ADVANCED DESIGNS FOR DISTRIBUTED COMPUTING

Within the last decade, a considerable amount of research has gone on into the best way to represent the architectural issues of distributed computing and also into the implementation of purist and pragmatic solutions. It is true to say that the final, definitive play-off between these solutions has not been resolved (and will probably be as much a marketing as a technical superiority result). However, it is also true to say that the different solutions appear to be migrating towards a high degree of consensus, or at least interoperability. Bearing these points in mind, we shall look briefly at one of the leading contenders, which embodies well the current thinking.

'COMMON OBJECT REQUEST BROKER ARCHITECTURE' – CORBA

CORBA is based on the programming concept of 'object oriented' design. That is, we first begin by defining the things we want to manipulate, their properties and the way they communicate with each other, in high-level terms rather than in terms of data structures and the way that computers deal with the data. A 'thing' could be a printer, a customer, etc.; obvious properties include names and addresses. Individual things can belong to a common group, a 'class', and share common properties, e.g. women and men belong to the people class and share one of its properties – an age. Object oriented design is not inherent to distributed processing, but it is extremely convenient, because it allows us to retain the essential concepts, in isolation from incidental and transitory events: customers are customers and they retain their particular properties whether their purchases are read by a point-of-sale terminal in a shop or made via a home computer. On the other hand, 'branch discount' is a property of the means of purchase, not the particular customer who purchases it. In a distributed environment, we often want to hide the geographical location of a service, because it is not material to the requirement and geography is a distraction; often we do not want to distinguish between local and remote processing or even whether a process runs on one machine or many. Object oriented design allows us to concentrate on the inherent properties.

The objects it uses are not just 'things' in the physical sense. Frequently objects represent processes, such as a 'value added tax gatherer' (TAXG). Applications programmers who want to use TAXG do not need to know how it works, but they do need to know some things (Figure 2.22).

TAXG is a 'black box' that is wholly defined by its definition of its service, which it makes public to applications, the way that applications should send messages to it, a set of details on how communication is handled that the applications do not need to know but that are there to inform the underlying COBRA infrastructure, and, finally, a unique identifier that will be used to locate it, even if it moves from place to place. (It is possible that an object that is calculating your tax can move from one computer to another. You still want to access that one, not one that is doing someone else's tax or doing the calculation on a different basis.)

A key feature of CORBA is the defining of objects in terms of these

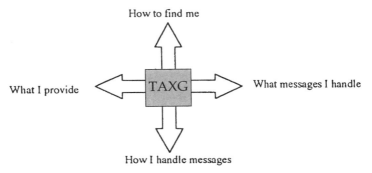

Figure 2.22 A CORBA tax-gatherer

four 'interfaces' to other objects and the infrastructure. These interfaces are specified in a specially designed Interface Definition Language (IDL). IDL is not intended to be *the* language in which all programs for distributed computing are written; instead it is constructed so that the components of IDL can be mapped onto other languages, using automated procedures. This means that programmers can write applications in languages of their choice and, assisted by the IDL manuals and tools, import into their programs the necessary interface 'functions' (sections of code that carry out specific routines – 'print', 'read', 'write', etc.) from IDL, without needing to know the intricacies of IDL. Because the functions, in any of the languages, always convert into exactly the same set of IDL instructions, programmers can be assured that the program will run on the CORBA-compliant system, irrespective of the source language.

'OBJECT REQUEST BROKER' – ORB

Imagine the case of a request from a 'client' (that is, a program somewhere in the distributed system) for an object. For example, the client may wish to invoke a TAXM object to deal with a tax calculation. In CORBA implementations, the client never communicates directly with the object; the request always passes via the Object Request Broker (ORB), part of the CORBA software that acts as a buffer between the client and the object (Figure 2.23).

The purpose of the ORB is to hide from the application programmer the complexities of dealing with a distributed environment, one which moreover may comprise a heterogeneous set of computers. The ORB finds the object and handles the detailed parameters of the

NEW NETWORK ARCHITECTURES

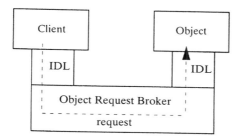

Figure 2.23 CORBA object request broker

request. Its task is, essentially, to provide 'location' and 'access' transparency. It does so through IDL interfaces as shown. (Incidentally, it can be seen that the ORB fits our definition of a CORBA object.) Within a distributed system, there may be several ORBs, each mounted on a different set of computers. Clients send their requests to their local ORB. Suppose the wanted object is not in the area covered by the local ORB? No problem: the local ORB simply communicates with the object's local ORB as in Figure 2.24.

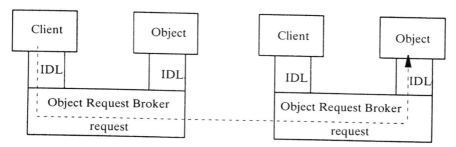

Figure 2.24 Distributed communication via other ORBs

A further advantage of this approach is that the different ORBs can reside on different vendors' hardware and software platforms and be written in completely different languages, to different internal designs. All that is required is that they are capable of providing the CORBA functions specified in their interfaces.

'OBJECT MANAGEMENT ARCHITECTURE' – OMA

CORBA itself is concerned with the interfaces between objects and how they are specified in IDL' this is the lowest level of a comprehensive

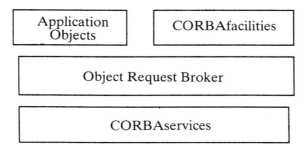

Figure 2.25 CORBA facilities and services

object management architecture (OMA) that comprises the vision of a non-profit-making cooperative group of vendors known as the 'Object Management Group'. Their idea is to create, by consensus, a common architecture for distributed computing, built round the object oriented principles. They separate basic, lower-level services for objects, the 'CORBAservices' level, from the facilities provided to applications, the 'CORBAfacilities' level (Figure 2.25).

CORBAservices comprise the more basic and generic needs common to most objects: things such as naming, querying, the grouping together of objects, security, and so on. A number of them are available in stable form from several vendors.

The CORBAfacilities are rather less well developed. Four basic categories have been defined: user interface, information management, systems management, and task management. These are clearly of use across all distributed business applications. 'Vertical' CORBAfacilities are also being generated for specific market sectors, such as healthcare and finance. The policy of the OMG is to adopt the existing models of information and processes already created by these sectors and integrate them into OMA, for example through creating new IDL interfaces.

CORBA NAMING AND TRADING SERVICES

From our perspective, probably the most significant of the CORBAservices are the 'naming' and 'trading' services. Naming is the simpler; CORBA has a set of system directories that set up a correspondence between the name of an object and its current location, just as a cellular telephone network can direct a call, described by a phone number, to a specific aerial serving the cell in which the called party is currently present. (You simply use the name; the CORBA service maintains the path to the object.)

But that assumes you know who you want to call (and also, possibly, for what purpose and on what terms). Suppose you want to find out who can sell you a particular type of car, at what price and how soon. You need to look at a Yellow Pages directory that lists companies under the services you offer, and obtain a contact number. That is part of what a Trader service will do. It will allow you to identify sources of objects you require; these may, in the early days, be simple services such as an application to securely 'watermark' your data against copyright theft, but, in principle, it should be able to find you a list of virtual organisations that can remotely run human resource management software for your organisation, for example.

The Trader not only gives you the names of the suitable objects; it also gives you details of the services they offer – speed, cost, security policy and so on.

At this point it has to be said that some of the Trader properties have still to be defined by the OMG and most of the benefits of Trader services can only be postulated. Nevertheless, the basic concepts are widely known and CORBA is becoming widely used as a platform for advanced services.

CORBA LIMITATIONS AND ALTERNATIVES

CORBA is not the only solution proposed for distributed computing. Others such as the Distributed Computing Environment (DCE) from the Open Software Foundation approach the problem from slightly different angles. DCE, for instance, has a client–server approach. However, CORBA and the credible alternatives probably have a reasonably smooth migration path towards a common solution, given industry willingness. The issues that the computing industry tends to concentrate on are those of increasing functionality without incurring equivalent costs in application software development or complexity, and in meeting the need to operate across a heterogeneous collection of hardware and software. This is a rational approach, given that these are major problem areas, but there is always a danger that other important issues may not be addressed. For instance, CORBA does not yet really support the management of multimedia data streams, and relatively little consideration is given by the various architectures to the role of low latency (end-to-end delay). The control of data across the network is, in general, very 'peripheral-centric'. Performance and quality of service are not yet the primary concerns of the computer industry's approach to distributed computing.

TELECOMMUNICATIONS 'SERVICE' ARCHITECTURES – TINA

Telecommunications organisations, on the other hand, whilst frequently criticised for their ponderous decision-making and the slowness of the rollout of new systems, can justifiably be proud of their comparative excellence in commitment to service quality and performance. Thus it is no surprise that they are concerned to construct network architectures from a basis of 'service'. The latest approach is a 'Telecommunications Networking Information Architecture' (TINA), driven by a consortium of telecommunications operators and equipment vendors (known as TINA-C). TINA is wider than the Intelligent Network concept described earlier; in particular, it recognises that the end-user and network-centred services can no longer be easily separated, nor can the network expect to control the specification and supply of peripheral equipment. Furthermore, with the ever-increasing complexity of services demanded of the network, it is not possible to decouple network and service management. For instance, if bit-rates can be purchased on demand, is this something that needs to be handled by a commercial, 'service-management' system that bills for it immediately it commences, or an issue for a 'network management' computer that enables the bit-rate provision? Clearly, it is an issue for them both.

TINA therefore recognises the need to consider things in terms of three different but interrelated architectures (Figure 2.26).

The approach is object-oriented, based heavily on the theoretical work of the International Organisation for Standardisation's Reference Model for Open Distributed Processing (ISO/RM-ODP), which also had a major influence on CORBA. The computational domain of the TINA architecture chooses to use the OMG IDL specification language, in common with CORBA. However, TINA lays much more emphasis on the 'stream interfaces' that handle 'isochronous' (that is, really real-time) flows of multimedia signals.

TINA is also much more detailed on the specification of network (that is, *telecommunication* network) resources and defines common classes of objects used by network services such as basic telephony,

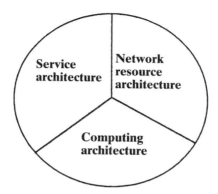

Figure 2.26 TINA

mobile telephony, multipoint multimedia and so on, irrespective of the underlying switching and transmission technologies.

TINA is concerned with a specific aspect of distributed computing and transmission, that of providing a flexible, secure, reliable telecommunications service, with tightly defined quality of service. As such, it exhibits some difference of emphasis from that of general-purpose distributed computing. However, it is encouraging to see that the two 'camps' of computing and telecommunications now appear to be talking very much the same language and, by and large, setting themselves the same goals. As we said at the beginning of this section, the major challenge to be achieved if future networks are to deliver their true potential is in the integration of approaches of this nature.

SUMMARY AND CONCLUSIONS

We have covered all the layers of our simple architectural model and looked at the central control systems and those on the periphery. We have seen how telephony, distributed computing and entertainment services architectures share some common points and differences. It would be false to claim that a clear, simple and certain picture of the future has emerged, but there are a number of likely ways ahead. We have seen that the most probable technology for transmission and switching in the main network will be ATM, or something like it, but the reason for this and, at the same time, the cause of much of ATM's problems is limited bit-rate. Also, we have seen how a variety of expedients such as radio or ADSL may solve the problems of local loop performance, at least for one-way traffic. In the next chapter we shall look at how optical fibre may alleviate the problem considerably. If this is so, and we no longer need to compromise on real-time operation versus cost, then many of the opposing viewpoints of the telecommunications and computing industries will disappear too. We note encouraging signs that the two industries are sharing a common outlook on higher-layer architectures for distributed information services. This is particularly important for the control-intensive services such as mobile telephony, which we cover later. A further trend is the significant redistribution of control from the centre to the periphery, brought about by the exponential growth in computing power.

3

Advanced Optical Fibre Networks

THE NEED FOR FAST BIT TRANSMISSION – SOME SCENARIOS

Suppose every evening, one million people decide to select a film or TV programme of their choice, at a time of their choosing. We could imagine the programme material stored on stacks of disks that are connected to the network and called down, 'on demand'. Everyone must be supplied individually with a bit stream which will run at a minimum of 2 Mbit/s, if they are to receive high-quality images and sound. That is a demand for 2 million million bits every second. This is at least two orders of magnitude greater than today's total telephone traffic for the UK.

Consider another scenario: medical diagnosis. Modern medicine is impressive, but expensive and rapidly becoming even more so. The UK spends some 5% of GDP on healthcare, these costs having tripled in the last three decades. It is impossible to provide full healthcare just exactly where it is needed. But it is possible that the difficulty of access to specialist consultants – which previously led to a cost equivalent to six times their salary – can be solved by providing wideband connections between them and the medical team in the patient's local hospital. Initially, a simple X-ray is carried out. Even for this, the data consists of about 10 Mbytes per image, and several are needed. These are sent over the network to the remote consultant. She needs more information, a CAT scan, NMR, PET, ultrasound, endoscopy and so on. They all require more and more data to be transmitted and their resolution is increasing every year. The ultimate desire is to be able to examine 'slices' of the body continuously rotating through any angle and at resolutions of surgical precision. The standard methods of compressing images onto limited-speed networks are not the answer: smoothing and sharpening images and error concealment are only acceptable provided they do not create artefacts or hide real problems. In any case, the analysis equipment itself is very expensive and cannot always be located close to the patient. What is required is a cheap 'head-end' that is sufficient to capture the necessary information and transmit it, probably uncompressed and unanalysed, to the centre, with a backward channel to allow the consultant to control the scan.

There is no reason in principle why remote healthcare should be restricted to examination. Surgery no longer relies solely on 'hands-on' movement of a scalpel. For instance, robot surgeons have demonstrated a performance superior to that of humans, when drilling bone. Precision cutting can be achieved over a distance, under the control of an expert. But here there is a very real problem with data compression: it has an inherent need to process over several images, leading to a delay. In general, the greater the bit reduction, the longer the buffer of stored data. That is, the longer the delay between the act and its image at the far end! The implications of this cut deep into remote surgery, quite literally.

So we have accepted that bandwidth requirements for remote healthcare will require considerable care and there is a need to err on the side of caution. We also have to note that images are not sent just once. They may do the rounds of several experts in several locations. They have to be stored for a number of years (21 for newborn children in the UK) and reliably accessed. Undoubtedly any attempt to quantify the amount of data required will result in an underestimate.

WHY DO WE NEED OPTICAL NETWORKS?

We need optical networks because our requirement for information transmission grows without bounds and optical networks are the only means of providing it economically. Ever since the beginning of telephony, the market has been driving technology faster and faster to find the answer. Pairs of copper wires gave way, in the 1950s and 60s, to coaxial tubes, but even their capacity reaches a practical limit of a few hundred megabits in the main (inter-exchange) network, and a real crisis of capacity arises in the local network between individuals and their first exchange, where use of flexible coaxial cable or twisted copper pairs of wires restrict it to a few tens of megabits per second. The solution to this bottleneck has been the invention of low-loss optical fibre. Not only does fibre provide a solution without equal for long-distance transmission, its low cost and small size offer an option that can revolutionise the 'last mile' between the local telephone exchange and the customer, where high bit-rate delivery had always been problematic.

THE EVOLUTION AND REVOLUTION OF OPTICAL NETWORKS

Since the 1980s, the infiltration of optical fibre into telecommunications networks has become an established fact. There are millions of kilometres of fibre installed in the world today (at least 1.5 million in the UK and over 10 million in the USA), and fibre is the preferred option not only for all new inter-exchange connections, but also for multiple business lines and, in some cases, for street cable TV networks.

However, despite the major economic and reliability improvements brought about by the introduction of optical fibre technology, the change has been incremental rather than radical. Fibres take up less room and require less line-plant, but today they still carry signals that have essentially the same structure as those formerly carried on copper: they use digital pulse techniques designed to meet the limitations of copper cables; they require all 'intelligent' functions such as switching and even something as basic as amplification and regeneration (reshaping) of the digital pulses to be carried out in the electrical, rather than optical, domain. This 'copper mind-set', as it has been described, abandons an enormous amount of the potential that could be realised from an (almost) 'all-optical network' that is based on its own capabilities rather than the limitations of its predecessors. It is instructive to speculate on the implications if we were able to design a radically new network architecture based on the properties of fibre alone.

Centralised Switching

In Chapter 2, we explained the existence of telephone exchanges as a way of avoiding the need to connect everyone to everyone else by a separate pair of wires, leading to around one billion pairs of wires running into each house in a country the size of Great Britain. This architectural constraint is the reason why, despite the dramatic change in the volume and nature of traffic, there has been a 'magic number' of 600 telephone exchanges for many decades. The enormous carrying capacity of fibre means that this constraint can be removed and we can begin to think of reducing significantly the number of exchanges, as well as greatly simplifying the way they operate.

Time-domain Multiplexing

It is possible to multiplex several individual channels onto one pair of wires, coaxial cable or optical fibre. Early methods of multiplexing several speech signals onto one cable relied on 'frequency division multiplexing' (explained later in this chapter). The development of cheap and reliable digital circuitry in the 1960s and 70s led to the demise of frequency division multiplex in favour of 'time division multiplex', where the speech signals were digitised and the bytes from multiple channels transmitted in time sequence, one after the other (see the frame on time division multiplexing in Chapter 2). There is a problem with time division multiplexing: suppose we have N calls, each of 64 Kbit/s, on a single cable coming into a switch and we wish to switch them to different outlets, for onward transmission. Then the switch must be able to make a switching decision at the aggregate bit-rate of $N \times 64$ Kbit/s. Thus, the ultimate speed of operation of a communication system may become limited by the achievable speed of the switch, not by the transmission cable. Currently, affordable digital electronics reaches its maximum at a few Gbit/s. As the demand for higher rates increases, we have to look for alternatives to the copper-based time division multiplex principle. Again, the capacity of fibre comes to our rescue, allowing a simpler switching structure with lower speed requirements.

Capacity Problems

We have explained that networks of the future will have to provide a multiservice capability: that is, they will need to handle voice and image with a guaranteed minimum bit-rate and a limit to the maximum end-to-end delay, at the same time as being able to cope with bursty data. To achieve all of this on limited capacity networks has required the construction of numerous transmission protocols, many of which are incompatible or can interwork only with difficulty. To make the transmission networks affordable, we have seen how telephony and computing data must be squeezed into the system using statistical multiplexing techniques such as ATM.

But, however we try, whenever the maximum capacity of the channel is lower than the theoretical sum of the individual data streams, sooner or later there is going to be unsatisfied demand. The theoretical carrying capacity of fibre removes this problem, at least on national cable routes. (Granted there will always be some areas where this is a problem, for example with already installed copper cables, satellite or local wireless systems, but it would be a pity to spoil the attractive simplicity of ATM, where it was not necessary.)

Electronics

Today's optical fibre systems are more truly 'electronic' than 'optical'. All multiplexing, amplification and switching is done by operating on electrical signals, either before they are converted to light, or after they have been converted back to electricity, or even by conversion from optical to electrical and back again, in mid-path. Obviously, this is expensive and potentially unreliable. There are special problems in hostile environments, particularly the local loop between the customer and the first exchange. Here there is an additional problem: how to provide reliable power supplies to drive any complex equipment that must be housed in a street cabinet or an underground cable chamber.

A NEW ARCHITECTURE

If we were to consider throwing away all the architectural constraints of a network architecture which has its roots in the properties and limitations of copper, and build a network based on the full potential of optical technology, in what way would it be different? We look first at the traditional, copper-based network, as shown in Figure 3.1.

Consider what is required in communicating between A and G: A to B and F to G traverse rather low-quality copper pairs of wires dedicated to the individual customer. From B to C (a main switching centre) the path may well be optical fibre, but 'playing by copper rules': for instance, the signal from A is first electrically combined with other signals, using time division multiplex. At C, D and E the signal is again converted to electrical form and switched electronically

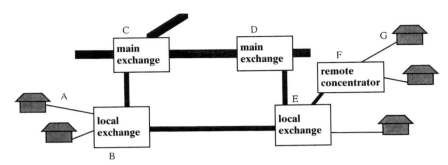

Figure 3.1 'Traditional' network

to the correct path, over a cable which is probably fibre. Bit-rate bottlenecks occur at various points in the network: there are customer to local network connections at a few hundred Kbit/s, unless special cable or transmission techniques are used; there is a switching bottleneck at the various exchanges; and there is a possibility of temporary loss of service at links served by statistical multiplexors.

Compare this with the case of the 'ultimate' optical network (Figure 3.2). A is provided with a full bit-rate, very fast, instantly available connection to B, C and D across their local optical networks and across an optical backbone without any tapering of the transmission capability, without any other loss of performance or complicated switching. The network becomes closer to a bus or a ring than to a tree with branches and roots. That is its logical form. Physically, it may be constructed quite differently, and in any case what we have shown is the ideal, ultimate, all-optical network, for which the technology is not yet quite there. But in the sections that follow, we shall show that realisable networks are coming quite close to that ideal. They retain some elements of the past: there is still a need for some centralised switching and we still make use of time division multiplexing, where it does not restrict performance; but, overall, we shall see how the prospect of virtually infinite carrying capacity sets us free from nearly all of the constraints.

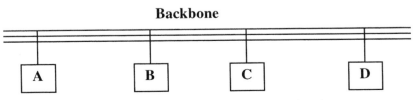

Figure 3.2 The 'ultimate' optical network

OPTICAL DIMENSIONS

Although the same basic building blocks are used in all communications systems, the way they operate depends on the technology deployed. Optical devices obviously operate using 'light'. Light is an electromagnetic wave phenomenon, just like wireless and, for that matter, electrical signals in coaxial cables and wires. We know that waves spreading out from a point are able to bend round obstacles in their path and interfere with other waves to build up into larger amplitudes or to cancel each other out. Our everyday experience teaches us, however, that light, wireless signals and basic voice telephone wires each behave in very different ways. Telephone wires do not radiate our telephone conversations into adjoining pairs of wires. That is, they are poor aerials, whereas quite short wires can be used for transmitting radio or TV signals. Radio signals bend round buildings but light appears to travel in straight lines. Optical fibres do not experience any induced noise. These differences come about because the wavelengths of each of their waves are of very different size. Consider the basic relationship connecting wavelength, velocity of propagation and frequency, shown in Figure 3.3.

The velocity of light in vacuum, copper cables or fibre is quire similar in all cases. In vacuum it is 300 000 000 m/s. In optical fibre it is about 50% less, and it is about one-third as great in copper cable. Let us look at the wavelengths of typical signals in each of these media.

Speech, directly converted to electrical signals by a telephone microphone, has a significant frequency content of around 3000 Hz, if treated thereafter as an analogue signal, or about 60 000 Hz if converted to digital form. (See the Appendix, Chapter 11, for the justification for this figure.) Therefore, the wavelength of speech signals

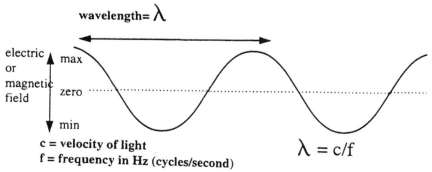

Figure 3.3 Basic parameters of a wave

carried in this direct form over a copper cable is of the order of 1 km (100 000 000/60 000 m).

In radio systems, the 60 000 Hz bandwidth is of little relevance to the behaviour of the radio wave, because the speech is 'modulated' onto a very high frequency 'carrier' (see frame). The highest carrier frequencies used in free space transmission between radio towers or from satellites are of the order of 30 GHz (30 000 000 000 Hz). This corresponds to a wavelength of 300 000 000/30 000 000 000 = 1/100 m.

Optical fibre systems use modulated beams of light to carry the signal. The light has a frequency of the order of 300 000 000 000 000 Hz, corresponding to a wavelength of 1/1 000 000 of a metre, usually referred to as 1 micron. (In fact, because the frequency is so high, it is usually more convenient to discuss optical systems in terms of their wavelength.)

MODULATION

It is sometimes convenient to convert the signal containing speech, video, data, etc., from its original form, the so-called 'baseband', into one that still contains the original message but has different transmission properties – for instance, using a high-frequency radio channel – or even a different means of transmission – as in the case of electrical signals from a microphone being converted to optical signals for transmission down a fibre. One way to do this is to take the original data and 'modulate' it onto a 'carrier' signal. The carrier does not contain any information; its function is simply to transport the original signal in a more efficient manner. The carrier is removed at the receiving end, in a process called 'demodulation', leaving the 'baseband' data. Two simple modulation schemes, 'amplitude modulation' which changes the size of the carrier depending on the value of the data, and 'frequency modulation' where the frequency of the carrier is altered, are shown here.

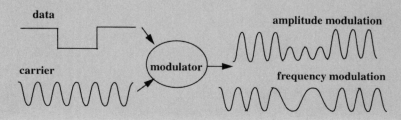

So, we see that optical wavelengths are of the order of 1/10 000 of the size of the highest usable radio waves and 1/1 000 000 000 that of the wavelengths of the original, 'baseband' signal. Two very important consequences result from this:

- Wave phenomena occur at a much smaller scale – the basic dimension that determines whether waves will bend round obstacles, interfere with each other, penetrate the surface of materials, and so on, depends not on the absolute distance in metres but on the ratio of the distance to the wavelength. An aerial must be a significant fraction of a wavelength in order to radiate well – that is why baseband, twisted pair, speech cables (each of whose twists creates aerials of a fraction of a metre) do not interfere significantly with each other. Long-wave wireless can 'bend' round tall buildings and even mountains, when the aerial is out of the line of sight. On the other hand, cellular radio and satellite systems, operating at GHz frequencies, make use of the fact that their area of operation *is* effectively line-of-sight; their signals can be obscured by obstacles a metre or so in dimension. To an optical system, a metre is a million wavelengths; optics concerns itself with dimensions of the order of the track widths of silicon integrated circuits. Optical components have 'aerials' that are a few microns in size. Beyond this distance, light does not really behave like a wave; it travels in 'rays', or at least we can frequently use the ray as a simple but adequate way of analysing optical systems.
- Channel capacity is related to carrier frequency – we discussed how the basic data we wish to transmit is frequently 'modulated' onto some form of carrier. The carrier is thus deviated from its normal value by the data. Clearly, the rate of variation of the carrier is directly related to the data-rate of the basic information modulated onto it. For some fairly fundamental reasons, it is a reasonable approximation to say that the achievable rate of variation of the carrier is proportional to the carrier's normal frequency. That is, *the data-rate achievable on any carrier system is proportional to the carrier frequency*. This is of obvious significance when we realise that the optical 'carrier', that is, light, has a frequency that is of the order of 10 000 times that of the highest usable radio frequencies.

OPTICAL COMPONENTS

The component parts of any telecommunications system are, at basics, similar irrespective of the technology used:

- We must have some medium for transmitting the signal without too much loss, over the required distance. In the optical case, this is an *optical fibre*.
- We need devices at either end that can generate and receive the signal, converting to and from the mode used for transmission and that required locally (for instance, sound waves to electricity). In the optical case, these are respectively *laser diodies* and *photodiodes*.
- There must be a way of routing the signal through the network so that it goes to the right place and does not interfere elsewhere with signals destined for other receivers. Optical systems will make use of components such as *optical splitters, couplers* and *filters*.

We now consider these elements in more detail.

OPTICAL FIBRES

The basic principle of optical fibre transmission is very simple, as can be seen from Figure 3.4.

If we have an interface between two transparent media, such as two glasses, which have different refractive indices (i.e., light travels at different velocities in them) and a ray of light passes from the medium with the higher refractive index, n_1, to the lower, n_2 (i.e. from the lower velocity to the higher), then the ray in n_2 will emerge closer to the interface than it was in n_1. If we make the angle in n_1 sufficiently

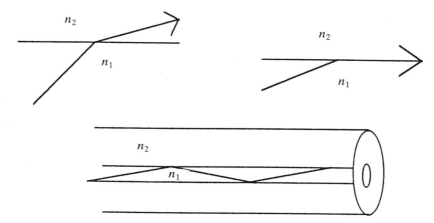

Figure 3.4 Principles of fibre transmission

small, then the ray will be bent so far back to the interface that it will never really pass into n_2. Making it even smaller means it will be totally internally reflected within n_1 at the same angle as it hit the interface. Optical fibres possess this dual refractive index and have a core (index n_1) which is so fine, typically of the order of 10 nm across, that any light launched into the core, by a laser or light-emitting diode, will enter at such a small angle that internal reflection will always occur.

If we combine this dimensional property with the additional requirement that the glass be manufactured in such a way that it is chemically very pure and free from imperfections, then the glass itself will not absorb much of the light and we can also get a fibre with very low loss of light over long distances.

FIBRE PERFORMANCE: LOSS AND BANDWIDTH

Figure 3.5 puts some values to the attenuation and wavelength range for fibre. The smoothness of the curve hides the dramatic progress that has been made over the past two decades in moving the overall curve downwards and in removing some major 'bumps' in it, particularly at 0.9 and 1.4 micron, where resonant absorption can occur because of water molecules trapped in the glass during manufacturing.

Converting this loss to a range in kilometres is rather complex, involving as it does a number of parameters: operating wavelength,

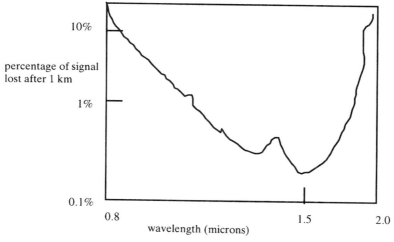

Figure 3.5 Attenuation per kilometre in optical fibre

bit-rate, performance of optical transmitters and receivers and, of course, the fact that long cables will inevitably require splicing, with significant fractions of transmission loss. However, with current technologies, it is easily possible to send signals over distances of hundreds of kilometres, without any amplification. In any case, as we shall see, there are ways whereby we can extend this range indefinitely, without requiring to convert from an optical signal back to an electrical one.

In practice, not all of this range of wavelength will be used, for there are some more subtle issues of fibre performance and there are restricted optimum ranges for the semiconductor transmitters and receivers that operate at the ends of the fibre. There are three particularly good regions of operation centred on 0.85 micron, 1.3 micron and 1.5 micron. Around each of these regions is about 25 000 GHz of bandwidth (approximately equivalent to a potential transmission bit-rate of at least 25 000 Gbit/s. *That is much more than the total telephone traffic carried in the UK today*. The fastest metallic cable, a rigid and heavy coaxial cable, operates at only a few tens of MHz over distances of tens of kilometres. The entire radio spectrum available is somewhat less than 30 GHz. The sudden availability of 25 000 Gbit/s on a single fibre cannot be dismissed as a mere incremental improvement.

If that were not all, there are number of other bonuses: by its very nature, fibre provides a very reliable and noise-free medium. Unlike wire pairs or even coaxial cable, there is no chance for interfering signals to be impressed onto the legitimate one. We can design end-to-end systems on the assumption that the optical path will be error-free. This leads to considerable simplification in terminal equipment and also in reducing system delay brought about by the need to carry out error checking and, possibly, resending of data. At the operational level, there are advantages too: fibre is considerably lighter, thinner and more flexible than its copper rivals; cable technology now ensures that it can be protected against damage and easily jointed or terminated without introducing significant losses. Thus, of the optical components required to create a transmission system, we can say that fibre will not be the one that limits the performance.

TRANSMITTERS AND RECEIVERS

The detailed theory of optical transmitters and receivers is extensive and complex. Receiver design is particularly complicated, but the basic optical detectors do not impose, at least in principle, any critical constraints on the overall performance of an optical network. Transmitters are more of a problem and we shall give a brief outline of the issues involved.

Transmitters in large-scale networks are invariably solid-state lasers, made from semiconductor material. This ensures that they are physically robust, reliable and relatively insensitive to temperature. Laser operation is required in order to obtain sufficient power and to produce light of a sufficiently narrow spectrum.

The phenomenon of 'lasing' is a result of a quantum-mechanical effect resulting when energy of the right frequency has been injected into the material and raises some of its electrons to enhanced energy levels. Under these conditions, a photon of light of the correct wavelength passing through the material can interact with one of the electrons and cause it to give up its extra energy, which is released as a photon with the same wavelength as the incoming photon. Thus light is amplified.

Obviously the semiconducting material is transparent (since light must travel through it in order to react with the electrons) and some of the light will leak out the ends of the device. But suppose we arrange for the ends of the device to be treated so they behave like partial mirrors; then some of the light is reflected back into the bulk of the material, and again stimulates the electrons to emit light of precisely the same wavelength, all over again. The more this is repeated, that is, the better the reflectivity of the mirrors, the narrower (or 'purer') becomes the bandwidth of the light emitted.

At this point we must distinguish carefully between two aspects of the laser's performance: its speed of operation, and its ability to operate over a range of wavelengths.

Speed of Operation

This is a measure of the maximum bit-rate that the laser can transmit/ We might at first think that this was simply the speed at which the laser can be turned on and off, in response to the pattern of bits to be sent, but in fact this is not a very suitable mode of operation. We mentioned that the repeated passage of light back and forward between the mirrored ends of the laser was required in order to set up the condition for lasing in a narrow frequency range. Switching on and off the laser brutally upsets the purity of the light output. Instead, we have to choose a less disruptive mode of impressing the digital signal onto the laser.

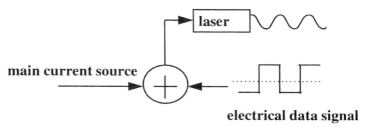

Figure 3.6 Modulation of laser by electric current

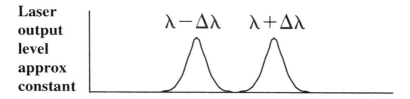

Optical wavelength varies with applied current

Figure 3.7 Output of current-modulated laser

One way, as shown in Figure 3.6, is to use the bit stream to vary the laser drive current by very much less than the difference between fully on and fully off. The effect is to vary the wavelength of the light emitted, while still keeping the lasing condition (Figure 3.7).

This variation in wavelength can be kept at a value as low as twice the bit-rate of the data stream, which is much less than the wavelength variation would have been if the laser were turned on and off. The variation is a tiny fraction of the unmodulated wavelength of the laser. There are a number of ways that this wavelength variation can be detected at the receiver and converted back into a bit stream. Rates in excess of 10 Gbit/s have been available for several years, using commercial components.

Now this figure of a few gigabits is several orders of magnitude lower than the capacity of a single optical fibre (which we said was about 25 000 Gbit/s at one of the chosen operating wavelengths) and, on its own, would represent a severe limitation to the usefulness of fibre, which brings us to the other aspect of laser performance.

Ability to Operate over a Range of Optical Wavelengths

We mentioned that lasers often possess mirrored ends. The mirrors, in effect, form a resonant cavity, and by altering the size of this cavity we

ADVANCED OPTICAL FIBRE NETWORKS

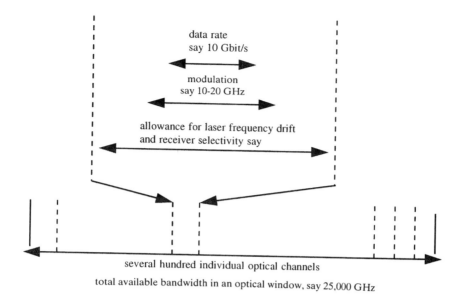

Figure 3.8 Capacity of a single fibre system

can alter the precise wavelength of the light emitted. By this means, or by other methods involving changing the dominant electron energy level or by altering the temperature, we can equip ourselves with a number of lasers operating at different wavelengths and use them over the same fibre.

As we can see from Figure 3.8, in practice we can space out the wavelengths of a few hundred lasers to allow them to operate without overlapping, in any one of the three 25 000 Gbit/s optical windows. (We would usually restrict ourselves to one of the windows.) Thus, as a ball-park figure for the best efforts to date, we can say:

one fibre system = 200 to 300 × 10 Gbit/s = 2000 Gbit/s

Purists will argue about this figure, mainly in the direction that it is too low, but the exact value is not important. What is important is the realisation that this is the performance of a complete and realisable *system*, not just the theoretical ability of fibre. This is a simple system that operates only in a 'point to point' mode without any means of switching from one optical channel to the next. However, it acts as the basic building block for more complex systems and, as such, is worth looking at in more detail.

POINT-TO-POINT WAVELENGTH DIVISION MULTIPLEX

As was said at the beginning of this chapter, there are millions of kilometres of optical fibre in service today, but they are almost exclusively making use of copper technology. In particular, they use time division multiplexing (TDM) to combine data channels, represented in electrical form, and then use this electrical signal to control a laser. An alternative would be to use a separate laser for each of the data channels, each operating at a different optical wavelength, and inject the separate optical signals onto the same fibre. At the other end, optical filters (see later) are used to direct the signal to the appropriate receiver. This technique, of carrying different signals separately over different wavelengths of light, is known as 'wavelength division multiplex'. Simple as it seems, it is the secret to low-cost, ultra-high-bandwidth, global networks.

Figure 3.9 shows the system in more detail, for the case of three separate data channels. (In general, there would be considerably more.) Each of the channels is assigned to a laser working at a different optical wavelength (w_1, w_2, w_3). The channels are combined onto one fibre using an optical wavelength division multiplexor, which we shall describe later, and demultiplexed at the far end in a similar manner. Note that a single fibre can carry signals in both directions simultaneously, although on long-haul links it is sometimes preferable to use two fibres, one for each direction, as this makes amplification easier.

TWO TYPES OF MULTIPLEXING

At this point, we ought to look carefully at our use of the term 'multiplexing', for it occurs in two very distinct ways: optical multiplexing and electrical multiplexing.

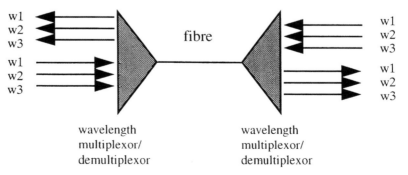

Figure 3.9 Wavelength division multiplexing

In the previous paragraphs we described how multiple channels could be provided on the same fibre by choosing lasers and filters that operated at different optical wavelengths. These we said could handle signals of several Gbit/s, and we could have several hundred simultaneous, independent channels each of this bandwidth. Because the independence of the channels is provided by purely optical separation on the basis of wavelength, this is the reason for the name 'wavelength division multiplex'. This term will be used only when the means of multiplexing is optical.

Of course, most times we may not have sources which require individual channels with bit rates of several Gbit/s; instead, we may have multiple channels of lower bit-rates, for example reasonably good-quality video links at 2 Mbit/s. It may be convenient to multiplex a number of these, electrically rather than optically, into streams that run at the maximum speed that can conveniently be handled by electronics, i.e. a few Gbit/s. We could do this using digital time division multiplexing, as shown in Figure 3.10.

Or we could use the perhaps simpler method of frequency division multiplexing, where the bit streams A, B and C are each modulated onto a different radio frequency carrier, which shifts their signals into non-overlapping parts of the radio spectrum, before the composite signal is converted from electrical form to optical, by the laser (Figure 3.11). Demultiplexing the signal at the other end is quite straightforward (Figure 3.12): the signal is converted from optical to electrical, filtered to select only the frequency band of interest and then each band demodulated using conventional radio techniques.

Frequency division multiplexing is particularly interesting, because it was the predecessor to time division multiplexing on copper cables but gave way to the latter because, among other reasons, the noisiness

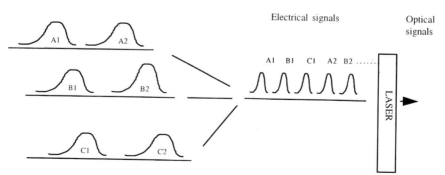

Figure 3.10 (Electrical) digital time division multiplexing

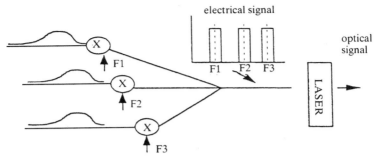

Figure 3.11   (Electrical) frequency division multiplexing

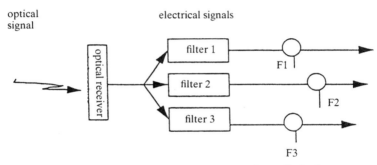

Figure 3.12   (Electrical) frequency demultiplexing

and long-distance attenuation of copper made it difficult to maintain signal quality with FDM. These are not problems with fibre and, maybe, there will be a resurgence of this technique.

OPTICAL COUPLING

We mentioned that the light from the three lasers was 'coupled' into the fibre. There is a need for another class of optical devices, 'couplers'. These can be formed in a number of ways, but most of them are variations on the theme of 'evanescent field' devices: it is only an approximation to say that light is entirely carried by total internal reflection within the 'core' (the glass with index n_1). In fact, a more detailed analysis of the propagation proves that there exists a weak electromagnetic field that penetrates the surrounding 'cladding, dropping off exponentially with distance away from the interface. Imagine a water wave passing along a trough which his thick

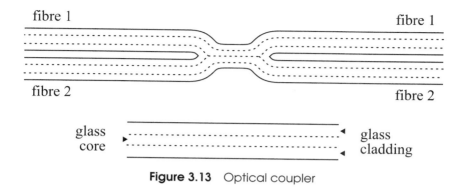

Figure 3.13 Optical coupler

'spongy' sides. The force of the waves will displace the elastic sides of the trough sideways, compressing the sponge when at a crest and releasing it at a trough. The wave does work in compressing the sponge, but gets energy back during the relaxing phase; thus no energy is lost at the sides. Similarly, the optical evanescent field does not result in loss of light from the fibre; it does not radiate out but rather follows the main light in the core along the length of the fibre.

This 'evanescent field' (so called because of the way it rapidly dies out with distance from the interface) is important, however: if we file away part of the cladding of two fibres and press them together, then their evanescent fields can overlap and transfer light from one fibre to the other. That is, we have made an 'optical coupler' (Figure 3.13).

This is exactly how one form of optical coupler is made: two fibres are deformed in some way, either by filing and gluing or by softening them in a flame and stretching and fusing them together. An alternative is to use standard photolithography techniques, originally developed for very large scale integrated circuits, to lay down closely spaced optical tracks which fields can interact in exactly the same way.

There is no reason why couplers can be restricted to two in, two out and, by altering the geometry of the fusion zone, we can alter the percentage of light coupled into each fibre. We can cascade couplers to allow one transmitter to send to a number of receivers or combine the output of a number of transmitters into one receiver fibre.

FIBRE TO THE HOME – PASSIVE OPTICAL NETWORKS, 'PONs'

Cascades of couplers can be used to create quite complex network architectures. Among these is the so-called 'passive optical network'

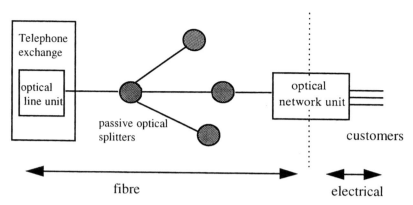

Figure 3.14 Passive optical network ('PON')

or PON: it is not just the major national and international routes that benefit from the introduction of fibre; the short hops of only a few kilometres that separate the vast majority of customers from their local exchange also require some radical, simple technology that can deliver reliable, broadband and *inexpensive* service. Complicated hybrid systems involving mixtures of electronics and optics in the access network between customers' premises and the exchange or concentrator have difficulty in meeting either the cost or the reliability constraints. One solution is an all-optical one, the 'passive optical network', PON for short. Figure 3.14 shows the basic principle.

The electrical signals for hundreds or even thousands of customers within a local area are electrically combined at the local exchange into one signal stream and then converted to an optical signal and injected into one fibre. The signal is split using a passive splitter based on the coupler principles we have described. The signal is distributed out through the network, eventually reaching a number of optical network units. These are either one-per-customer or, more likely in the early days at least (because of cost), one per group of several customers served by the same street cabinet. The optical signals can then be converted back to electrical signals and delivered to the home over short lengths of existing copper cable, if the cost of direct fibre to the home has not yet dropped. At present it is more economical for each optical network unit to serve around 5–20 customers.

Notice that all optical network units receive the same composite signal, containing all the channels. To provide security and privacy, the signal therefore has to be encrypted, with decryption units provided for each individual channel.

A system of this type is very convenient for distributing the same wideband signal to all premises, for example for multichannel TV. However, we have to consider the question of how return signals are sent over the system, where we wish to provide both-ways interaction. There are number of possible solutions. We can transmit the downstream signal in bursts, leaving a silent period during which return signals can be returned in sequence from the optical network units (with the network units synchronised via a network clock); another way is to use two different electrical frequency bands for the 'go' and 'return' signals; yet again, we can use two different optical wavelengths.

OPTICAL FILTERS

The passive optical network can cope with a large number of customers within a metropolitan area, but we are still not in possession of all the components required to create a truly global network. So far, for instance, we have been restricted to electrical switching or steering of channels. We need components that allow us to steer different optical channels selectively. Pure coupling is not particularly selective; it lets light of any wavelength couple across the device. Sometimes this is not a good thing; we may be using light of one wavelength for one purpose and light of another for another. Here we require the optical equivalent of the electronic filter.

We touched briefly upon the subject of tuning when we discussed lasers. In particular, we mentioned that, by converting the ends of the laser into mirrors, we could selectively tune the laser's wavelength. This is an example of a particular form of filter, the 'single cavity Fabry-Perot Interferometer' which simply consists of a length of transparent material (e.g. air or fibre) between two half-silvered mirrors (Figure 3.15).

The right-hand graph shows the intensity of the light passing through the system, as a function of the wavelength, the gap width x, and the degree of silvering of the mirrors (for which three different values are shown). We can see that, simple by changing x, we can alter the amount of light at any wavelength passing through the system. We can tune the filter to give us a maximum intensity of light at one wavelength and a minimum at another.

The gap, x, could be controlled by a simple vernier screw mechanism, hand- or motor-driven, but it is more usual to use an electrical

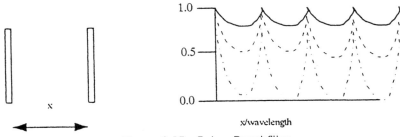

Figure 3.15 Fabry-Perot filter

technique, such as a piezo-electric material connecting the two mirrors together, which can be arranged to widen or shorten the gap in response to an impressed voltage. There is not necessarily a need for discrete mirrors; two fibres with their ends polished and spaced apart using a piezo-electric sleeve can do the trick. One advantage of this electrical control is that feedback can be automatically introduced to compensate for such things as thermal expansion, which otherwise would alter the gap size. (The gap size is measured electrically, for instance by measuring electrical capacity across the gap – this is inversely related to gap width – and a compensating voltage applied to bring the size back to the correct value.)

A mechanical system like this, even when driven electrically, does have a slight limitation: speed of operation, which runs to several milliseconds. This is not as big a problem as it might seem. Although the bit-rates of signals passing through the filter can be many tens, or even hundreds, of gigabits per second, the filter does not have to tune at this rate. Remember, we have escaped the TDM bottleneck: the filter only needs to respond at a speed suitable for setting up the connection. However, there are occasions when we might wish for something faster, particularly if we are considering a packet-switched network rather than one where semi-permanent connections are set up.

Consequently there is much interest in designing filters which utilise direct influence of the electric field on the optical properties of the material (rather than going through the intermediate stage of electro to mechanical conversion, as in the example discussed above) and similarly, in acousto-optic tuneable filters, which make use of vibration (at MHz frequencies) within the optical material. In both cases, the applied energy, whether electrical or acoustic, affects the refractive index of the material, thus changing its effective optical length.

BUILDING THE ALL-OPTICAL NETWORK

We now have the majority of components we require in order to build very large optical networks. In the sections that follow, we are going to outline the design of a global optical network, based on wavelength division multiplexing and employing component elements whose principles are described above.

Our earlier example of a point-to-point wavelength division multiplex system had, as its name implies, the problem that we could not send the optical signals to any point other than that set up in the beginning. Figure 3.16 shows how we can get round this problem, by introducing optical couplers and tuneable filters, to create a 'broadcast-and-select' system. Each of the laser transmitters in the diagram is set to transmit at a fixed optical wavelength. The passive 'star' coupler broadcasts signals from each laser into all the other fibres. Thus each receiver receives signals from all of the transmitters. However, the receivers are equipped with tuneable wavelength filters as described earlier and they are able to reject every wavelength except the one they are set to select. Note that several receivers can choose to tune in to the same transmitter. This is advantageous in the case of multiparty, video-conferencing, or cable TV broadcasting, for instance.

Although we have shown fixed transmitters and tuneable receivers, it is also possible to have tuneable transmitters. If receivers are fixed and transmitters variable, then transmitters would have to tune to the appropriate wavelength corresponding to the receiver they wished to call; if the opposite were the case, then receivers would have to tune to listen in to each transmitter's wavelength to see if they were being

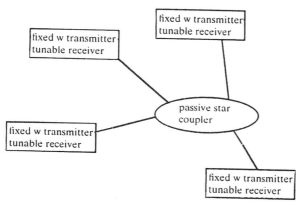

Figure 3.16 A 'broadcast-and-select' system

called; if both are variable, it becomes rather more complex. An alternative is to reserve a single wavelength to operate as a common signalling channel for all transmitters and equip every receiver with a wavelength filter that can receive this channel. We shall say more about this when we come to discuss large-scale networks. One important point to notice is that this control channel does not need to operate at anything like the fantastic speed that would be required for a TDM system. Like the telephone operator, it merely has to work at the speed required to set up the path.

NETWORKS WITH WAVELENGTH ROUTING

The broadcast-and-select network described above can provide a very useful way of distributing to an enormous number of receivers, down one fibre, but it suffers from the problem that we currently cannot put more than 100 or so laser sources within any one optical window. Granted the individual channels are wide, but the routing capability is extremely limited. However, we can get round this in a very simple way: rather than build a single, all-embracing network, we split it into a number of subnetworks that are, in general, isolated from each other, except when we require to pass selected wavebands between them. In Figure 3.17, we see two identical broadcast-and-select networks, joined together via a filter.

Suppose transmitter A, on the left-hand network, wishes to communicate with receiver B in the right-hand one. All that is necessary is for the filter to be tuned to pass the wavelength of A's laser whilst blocking off all the other wavelengths. Apart from this wavelength, the networks can function independently, setting up connections

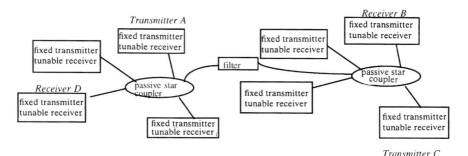

Figure 3.17 Wavelength-routing network

ADVANCED OPTICAL FIBRE NETWORKS

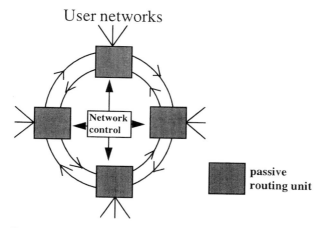

Figure 3.18 A wide area wavelength-routing network

between their own transmitters and receivers as before, using all the other wavelengths. That is, all of the wavelengths except for the one that bridges between the two networks can be reused. To set up another bridge between, say, transmitter C and receiver D, we simple select another wavelength that is not used by either of the individual networks.

Thus the secret is simple: wavelength reuse outside the local subnetwork. There are many configurations that allow us to do this. One example is given in Figure 3.18. Here the individual users are grouped together onto individual passive optical networks as described earlier. Each of the PONs is connected into a wider 'double ring' network by means of a 'passive routine unit'. A centralised network control assigns wavelengths to the passive routers in order to allow them to switch traffic between networks. A typical configuration of a passive routing unit is given in Figure 3.19.

It is not necessary to work out the detailed operation of this configuration to notice how the router is constructed from a number of adjustable filters and passive splitters and combiners. The network shown is a double ring, which was chosen to facilitate explanation. In practice, it is more efficient to use grid or mesh-shaped topologies which have more direct connections between nodes, but the principle is just the same. In fact, the double-ring configuration is not entirely of theoretical interest. It is an effective way of providing a 'metropolitan area network' (MAN) to serve a medium-sized city, for example. This may be a way for cable TV companies to make an early entry into the all-fibre provisioning of systems.

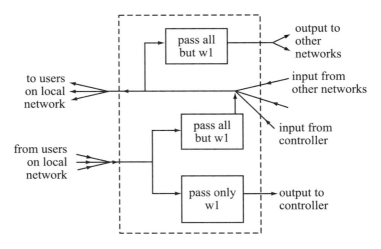

Figure 3.19 A passive optical routing node

THE CARRYING CAPACITY

Let us put the design into context by discussing the achievable capacity of such a system. It is important to understand that the system is fundamentally asymmetric. That is because the connections come via a passive optical network, as described above. As we saw, signals from the centre of the network fan out, via optical splitters, to the individual customers (or at least to local distribution points), whereas signals from the individual customers are aggregated together using optical combiners. This results in a difference in performance in the two directions when we consider the need for optical amplification, which is necessary in the case of long distances or multiple splits.

Look first at the 'downstream' case, where we broadcast from a single point, for example in the case of digital TV. It is already possible to demonstrate, in the laboratory, systems which serve tens of millions of customers with nearly 10 000 high-quality TV channels each over a distance in excess of 500 k. Theoretical calculations suggest it may be possible to increase this to 200 000 channels!

The situation is not quite so favourable in the 'upstream' direction. Because each customer's signal has added to it the summation of the amplification noise from all of the other signals combined onto the same fibre, it turns out that we reach a practical limit of around 100 000 customers at 5 Mbit/s each. This suggests that we need to introduce a degree of centralised switching into our system if we wish to cover a customer base as large as the UK, for example.

NATIONAL COVERAGE

One way to do this is to go to a highly connected 'star' configuration of optical networks, routing them to a centralised multiplexor (Figure 3.20).

We take the most radical option: each of the individual optical networks is arranged so that each individual customer has an all-optical connection, consisting of a tuneable optical receiver and transmitter. Several customers are combined together through optical combiners to give a composite upstream optical signal (and the downstream signals are optically split). The signals are optically amplified (we shall see how, in a later section) and then passed to the passive router, whose mode of operation is shown in Figure 3.21.

Light comes into the router via light guide A. This could be a fibre

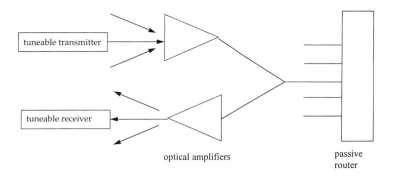

Figure 3.20 Centralised routing

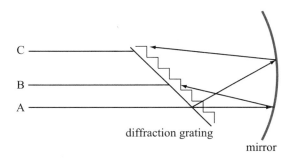

Figure 3.21 Passive optical router

but is more likely to be an optical channel in an easily machined optical material such as lithium niobate or a polymer, on which the fibre terminates. The light from the end of A spreads out (a microlens may be used) and impinges on the mirror. It is reflected back towards the diffraction grating. Because of the way diffraction gratings operate, light will pass through to the output paths B and C only if there is the right combination of wavelengths and angle of incidence on the grating. Thus it is possible that light will pass into guide B , but not into C. We could change this by changing the angle of some of the components, but there is an easier way and one which is much more precise – change the frequency of the light. Thus by selecting the transmitter and reception frequency centrally, we can channel the signal across the path between the two, in a fully optical and passive manner. (An even simpler option is to arrange the microlens such that broad beams of light from each input are injected into *all* of the return fibres, essentially producing a splitting function similar to fibre splitters but with a much bigger fan-out.)

OPTICAL AMPLIFICATION

Providing national and international networks is not just about combing and splitting signals, but also about transmitting them over long distances. Although optical fibres do possess very small optical loss, it is never zero; in any case, joints and couplers introduce additional lossy points and, of course, every time we introduce an optical splitter we divide up the optical power between the branches. This means that we need periodically to amplify the signal, to keep it above system noise (which is mainly developed in the receivers and any other active components, rather than the fibre). Traditionally this is achieved by first converting the optical signal to an electrical one, amplifying and reshaping this, and converting it back to optical. Obviously, this means breaking into the fibre, providing some form of enclosure for the electronic components and also supplying them with power. The electronic components have to operate at very high bit-rates and maintain low jitter regeneration of the signal. Whether the system is a short-haul terrestrial, local distribution one or an intercontinental submarine cable, this is inconvenient and increases the cost and, through its complexity and discontinuities, adversely affects reliability. That is why all-optical amplifiers offer an attractive alternative.

ADVANCED OPTICAL FIBRE NETWORKS

Over the last 10 years a number of optical amplifiers have been developed. The first successes were had with semiconductor elements which consisted of a semiconductor laser interposed in the fibre path. The strong and coherent light of the laser 'pumps' electrons within the semiconductor into excited energy states; when the data stream from the fibre passes through this area it causes some of the electrons to drop to a lower energy level with the consequence that they emit light in phase with the data pulses, thus producing an amplification. This laser still requires the presence of electrical power and some low-speed electronic components, but is a considerable simplification over an electronic amplifier.

To a good approximation, optical fibres are highly 'linear' devices – that is, their output is almost exactly proportional to any input, and where there is more than one input (e.g. from two laser transmitters) these inputs do not interact in the fibre. However, this is only an approximation and optical amplifiers have been developed that make use of the slight deviation from linearity. Even undoped fibre, when pumped with a high-powered laser beam, can provide amplification of a lower intensity signal-bearing beam due to the non-linear scattering process induced by molecular vibrations in the glass structure.

More significantly, there have been breakthroughs in the development of amplifiers which use the inherent properties of 'doped' fibre. One such example is shown in Figure 3.22.

The principle is, again, very simple: lengths of the order of a few tens of metres of fibre are doped with the rare-earth element erbium. These doped lengths are spliced into the normal transmission fibre, together with an optical coupler which is connected to a solid-state laser. This solid-state laser continuously provides optical power into the doped region. In this region, the light is absorbed by electrons

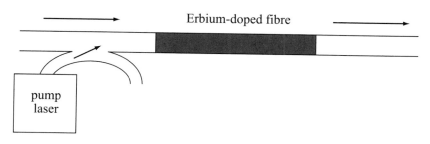

Figure 3.22 Erbium amplifier

which then become raised to higher energy states. When light from an incoming signal passes into the doped region, it can stimulate some of these electrons to drop back into lower states, releasing photons of light in phase with the signal. Amplification of the order of 20 times can be achieved, over wide optical bandwidths. Note that the pump laser runs continuously; it does not require high-speed electronics to drive it. That greatly simplifies the design, improves reliability and reduces cost, of any amplifier that has to be installed along the long-distance route, compared to an electronic solution. It is even feasible, in some instances, to locate the pump laser at some more convenient site, and feed its optical power to the erbium section, over an optical fibre.

Long-distance systems can incorporate several hundreds of fibre amplifiers without problems. In the next section, when we look at optical solitons, we shall see how fibre amplifiers are a key component in futuristic optical systems.

FURTHER AND FASTER

So far we have looked at optical fibre systems based on components that are available 'off the shelf', but a measure of the maturity of a technology is how much further improvement we can hope to make, once ideas, now in the laboratory, move into the field, and optical technology seems to promise a great deal to come. There are interesting developments, for instance, in 'coherent' optical systems, where the optical receivers operate on the same principle as today's radios, with the phase of the optical signal becoming important, rather than just the intensity of the light. There are optical wavelength converters which, rather than just filter out optical bands of specific wavelengths, actually change the wavelength of the light, thus introducing another degree of freedom for optical switching and other devices. At least from the viewpoint of intercontinental links, perhaps the most interesting optical phenomenon currently being incorporated into experimental systems is the 'soliton'.

SOLITONS

Suppose we had no problems in developing fast, stable laser transmitters and optical receivers. What would be the maximum limit to the bit-rate we could transmit down an optical fibre? It turns out that the answer to this question does not depend on the absolute speed at which we can transmit through the glass, but rather on relative speeds of the different components that go into making up the optical pulses. Remember that a pulse can be represented as the composite of a set of individual pure frequencies (or, in the optical terms we have been employing, their equivalent 'wavelengths') and the shorter and sharper the pulse, the bigger the spread of the wavelengths. (See the Appendix, Chapter 11, for a further explanation.) Thus projecting a short pulse of light into one end of a fibre is equivalent to projecting a series of pure wavelengths, starting them all off at the same time. But what happens at the other end? If the fibre is short, or if all wavelengths travel at the same velocity, there is no problem. However, the velocity of light in fibre varies sufficiently with wavelength that, over intercontinental distances, the pulse spreads out significantly as a consequence of the different wavelengths taking different times. The phenomenon of different velocity for different wavelength is called 'dispersion' and severely restricts the maximum achievable bit-rate.

It is possible to alter the composition of the glass in order to minimise this dispersion effect, but not to remove it completely. However, there is another property of glass which can be used to get round the problem: the non-linearity of the velocity of light in the fibre with respect to optical power. To a good approximation:

$$V = V_0 + N \times I$$

That is, the actual velocity of light of intensity I in the fibre is equal to a constant, V_0, plus a constant, N, multiplied by the intensity. Clearly, this difference in velocity between different parts of a pulse of varying amplitude will lead to a change in the shape of the pulse as it progresses through the fibre. But we have said previously that the dispersion of the fibre tends to alter the shape of the pulse also. Is it possible to oppose those two processes, so that the pulse maintains its original, compact shape? The mathematics are rather complex, but it turns out that this is possible, provided two conditions are met: the pulse must have a particular shape, depending on the dispersion and amplitude non-linearity of the fibre involved, and the amplitude of the pulse must not fall below a level whereby the amplitude non-linearity ceases to be significant.

The first condition (pulse shape) is relatively easy to meet; the second condition has also fortunately been made possible by the development of optical amplifiers, as mentioned earlier. Provided the pulse amplitude is maintained within certain bounds, by means of periodic amplification, it can be shown that these pulses, 'solitons' as they are known, can travel extremely long distances without losing shape. Experimental systems have achieved 10 Gbit/s, or more, over distances of 1200 km with virtually zero error rates.

CONCLUSIONS

The components described above can be used to create networks with radically different properties:

- Bandwidth: we offer the opportunity to remove the local loop bottleneck by offering, if required, gigabits per second to each individual customer.
- Multiplexing: these channels can be offered using conventional electronically multiplexed techniques, but without setting too stringent a demand on the speed of the multiplexing or switching equipment, because we can use a large number of lasers operating at different optical wavelengths. Alternatively, we could offer individual optical wavelengths to each customer, with gigahertz bandwidth. The choice of all-optical or optical with some electrical multiplexing will depend on the comparative cost at installation time. It is completely feasible to mix and match both systems.
- Public switches ('telephone exchanges') will be dramatically reduced in number, in principle to five or six for the whole of the UK.
- Dramatically lower costs: the cost of providing national trunk networks and switches can drop by at least one order of magnitude, yet the available bandwidth will go up by several orders of magnitude!

TIMESCALES

When will it all happen? Firstly, there are still some hurdles to overcome in the technology. Optical devices are still rather expensive and mass-production methods are available for only some of the components we describe above. Partly, this is because the technology does not, in general, lend itself so easily to the microlithographic processes that are used so successfully to manufacture silicon inte-

ADVANCED OPTICAL FIBRE NETWORKS 115

grated circuits, thus reducing costs by several orders of magnitude. But also, commitment to mass production, in the case of silicon and optics, must be preceded by a belief that a mass market exists, because of the high up-front costs. We can only form an estimate because the major investments that have to be made must be based on market judgements, which are much more difficult to prove than the technology. Here we suggest a possible scenario for optical network deployment, in terms of timescale and motivating factors. It is overwhelmingly probable that networks will 'go optical' in two or three decades, but whether they do so as part of a gradual evolution or as part of a revolution in a turbulent marketplace is impossible to predict (Table 3.1).

As we have said, there is already a sizeable investment in optical technology, albeit based on the copper paradigm. By the start of the next century, we expect a number of important users to be utilising optical networks which are significantly closer to the all-optical model we shall describe. Who these users are, it is difficult to say: they may be major companies with very large amounts of data to ship around, but they are just as likely to be providers of mass entertainment or education. Because of the investment in cable installation, the first markets may be for metropolitan, rather than global, networks, offering ultra-wideband connectivity across major cities, without any intermediate electronics.

Table 3.1 Possible evolution of fibre networks

Timetable for optical deployment	Implementation
Today: 'Optical copper'	Optical fibre trunk transmission of traditional time division multiplex, electronic switching, optical 'cable TV', point-to-point optical links
2000–2010: 'Age of optical evolution'	Fibre to the fortunate few, passive optical metropolitan networks
or 2005–: 'Optical revolution'	Green-field national networks are completely optical, with electrical signals only at edge, explosion of bandwidth, tariff wars
2010+	Adoption of fibre and fibre-based 'mind-set' soon dominates design

A variation of this would be aggressive competition between companies to provide services to selected large customers as part of a global network that bypasses the dominant national and local telecommunications companies. This could be less than a decade away. Again, 'tiger economies' in small, possibly centrally controlled, countries could command the installation of a green-field all-optical national network for competitive advantage. This could be very awkward for existing large telecommunications companies, leading to tariff wars that they would not like to contemplate.

4

Radical Options for Radio

SCENARIO: UNIVERSAL PERSONAL MOBILITY

Within the next decade, it may well be possible to travel anywhere in the world and still make telephone calls or send and receive data, without changing terminals or even having to seek out a fixed telephone point. There will be considerable compatibility between the services offered by the fixed and mobile networks (caller identity, voice messaging and 'follow-me' facilities and so on). In 15 years or less, advances in mobile communications will give us access, anywhere, to high-speed intercontinental channels through which we can provide or receive high-quality moving images and sound. Third-world countries will be provided with communication systems without the high cost of a complex infrastructure.

In our houses, we shall soon see a rapid blurring of the distinction between 'mobile' and 'fixed-line' telephony, and in our offices we shall observe the extensive introduction of 'wireless' local area networks that remove the complicated and messy wiring plans we have today.

A NEW AGE OF WIRELESS

During the 1960s and 70s, received wisdom said that telecommunications was about finding faster wires: copper pair had given way to coaxial cable and experiments were underway with helical waveguides that could handle the expected growth in trunk traffic by constraining the signals within copper pipes. There had been some experiments with local distribution of telephony and TV using GHz radio but, by and large, 'radio' meant expedient, expensive and noisy alternatives to trunk cables. Even satellite systems were seen as poor alternatives for submarine cables, a view that gained force with the developments in optical fibre.

All this was to change at the end of the 1970s and into the 80s with the coming of cellular telephony. The phenomenal take-up of mobile phones exceeded the most optimistic forecasts of the marketeers. (For example, the relatively recent introduction to Italy of cellular services has led to a demand that has exceeded expectation by 1000%!) The GSM digital cellular service that went live in 1992 continues to demonstrate this exponential growth, as shown in Figure 4.1.

What made this possible? After all, neither the amount of radio spectrum nor basic radio principles had changed (see the frame below). What was new, however, was the development of very large scale integrated silicon circuits, which continue to offer a doubling of performance against cost every 1–2 years, leading to a dramatic fall in the price of handsets, and the equally impressive introduction of software control for the tracking of the mobile handsets and the switching of their calls. To best understand what became possible, we shall first look at cellular networks, the principle behind today's 'mobile phone'.

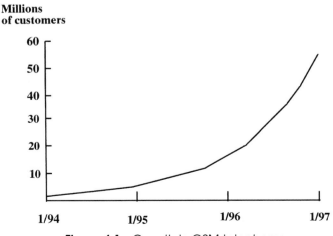

Figure 4.1 Growth in GSM telephony

RADIO SPECTRUM

Radio signals are electromagnetic waves, essentially the same as light waves but operating at much lower frequencies (see the section on 'optical dimensions' in Chapter 3). Although they are of lower frequency than light, they are generally of higher frequency than speech waveforms or the data-rates of moving video. The higher the frequency, the more effectively the signals can be radiated from antennae ('aerials'); that is why the speech or video is usually 'modulated' (see the frame in Chapter 3, 'Modulation') onto a radio frequency carrier. Bands of radio frequencies are designated somewhat differently around the world, but a common designation is as follows:

Frequency range (GHz)	Designation	Telecoms/broadcast usage
0.1–0.3	VHF	Terrestrial radio, TV, cellular, weather satellites
0.3–1.0	UHF	TV, GSM cellular at 0.9 GHz
1.0–2.0	L	Navigation and mobile satellite services, GSM at 1.8 GHz
2.0–4.0	S	
4.0–8.0	C	Fixed satellite services
8.0–12.0	X	
12–18	Ku	Direct broadcast services
18–24	K	Low earth orbit ('LEO') satellite-to-ground link 19–20 GHz, LEO satellite-to-satellite 23 GHz
24–40	Ka	LEO ground-to-satellite link 29 GHz. 38 GHz is used for local distribution within 8 km radius, e.g. for wireless fibre systems. Susceptible to attenuation by rain. The high frequency allows narrow beam forming and small (30–60 cm diameter) building-mounted dishes
40–100	mm	

PRINCIPLES OF CELLULAR NETWORKS

Central to the operation of virtually all current and future radio networks is the principle of *spectrum reuse*. There is only a limited amount of radio spectrum available, nothing like enough to satisfy our requirements if we were to allocate each customer a permanent band

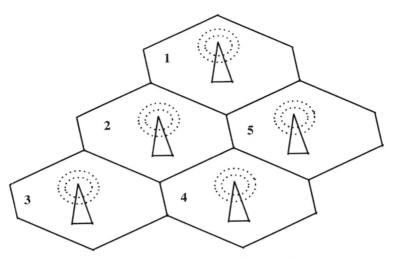

Figure 4.2 Principles of cellular radio

on which to broadcast and receive powerful signals. Instead, we have to do two things: firstly, allocate them a frequency only when they are actually making or receiving a call; secondly, severely limit the power and, at least in the case of the base-stations, restrict the direction of the transmissions. (Because the mobile handset is physically small, its aerial cannot be designed to have very directional properties; this is not, however, the case with the base-station transmitters.) In Figure 4.2, we see an idealised view of how this might be achieved.

The area to be covered by radio is split up into a number of contingent 'cells'. In the diagram they are shown as very regular in shape and size; in practice their dimensions would vary extensively, depending on the density of users and the natural and man-made topology (hills, tall buildings, etc.). Each cell contains a transmitter/receiver base-station. The base-stations can broadcast and receive a number of radio channels, but their transmitted power and the height of their aerials are designed to control the range over which they spread. Base-station 1, for example, can reach all mobile handsets within its own hexagon, and can receive transmissions from handsets in cells 2 and 5. It also transmits signals which, unfortunately, cannot practically be limited to its own cell. They are sufficiently strong to rule out the possibility of the same frequency bands being separately reused in cells 2 and 5. By the time its signals reach cells 3 and 4, however, their strength is so weak (either because of distance or

RADICAL OPTIONS FOR RADIO

because of obstruction by landscape features) that they would not create any serious interference, thus allowing the same frequencies to be reused in 3 or 4. Clearly, there is a close analogy with the reuse of the optical spectrum, as described in the previous chapter, but with the important distinction that, whereas we can rigidly constrain the light to lie strictly within any set of fibres, for free-space radio we must introduce a 'fire break' of dead ground to cope with the gradual die-off of the signal.

The actual transmission frequencies used to make calls by the mobile and the base-station are assigned 'on-demand' by the intelligent processing associated with the latter. Mobiles continually monitor across the spectrum, listening to common signalling channels from the base-stations, usually locking on to the strongest signal, corresponding to the cell they are in. This data channel alerts them to any incoming calls and also assigns them a pair of frequencies, one to receive the incoming caller and the other for speaking back. As mobiles move about, the strengths of the signals transmitted and received are constantly being monitored. Whenever the signal strength becomes too weak, the mobile and the base-stations negotiate a 'handover' (or 'hand-off' in the USA) to a stronger transmission path.

INTEGRATION WITH THE FIXED NETWORK

If we were to restrict calls within the mobile network to be between handsets both in the same cell, then we could easily design a simple switching subsystem, located at the base-station, that could set up the call. Of course, this is seldom the case, as a useful mobile system should provide national or even global coverage. Rather than set up a completely separate wide-area radio network, consuming radio spectrum and significant capital, it makes great sense to integrate the radio network with the existing fixed network, as shown in Figure 4.3.

In the figure, we see the individual base-stations connected to a 'mobile switching centre' which, confusingly, despite this standard term, is a fixed site. The connection can be either direct or chained via another base-station, and by cable or by radio. The mobile switching centre is connected to the fixed network. Obviously, the fixed network is used to carry the long-distance traffic. What is not quite so immediately obvious is its active role in controlling the routing of the call.

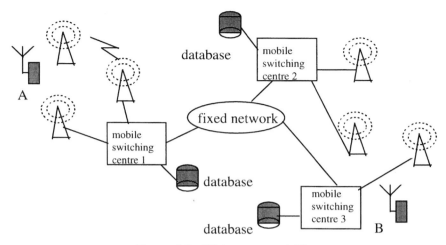

Figure 4.3 Wide area mobility

'ROAMING' AND THE ROUTING OF CALLS

By introducing a wide-area capability (through use of the fixed network), we see that calls can be connected between mobiles in widely separated cells. But how does the system know where to route the call, when the mobiles can 'roam' across many cells? As shown in Figure 4.3, each mobile switching centre is associated with a database. For each mobile handset, there is an entry on a master database, for example a database located in the region where the mobile was purchased. Suppose mobile handset B was originally registered on the database associated with mobile switching unit 2 and suppose mobile A wishes to contact mobile B. Provided mobile B has been left switched on, the following sequence occurs:

- B periodically locks onto the appropriate transmitter for the cell it is in and transmits a 'registration' signal, thus identifying its location (in the case of Figure 4.3, a cell within mobile switching unit 3's area).
- Mobile switching unit 3 passes the information on B's current location, through the network to the database of switching unit 2 (i.e. where B was originally registered). The information is stored there.
- Now let A dial B's number. The dialling information is processed at mobile unit 1, which finds B's details in its database and sees that B was originally registered on the database associated with switching unit 2.

- Unit 1 sends a request over the network to unit 2, asking for the current location of B.
- Unit 2 replies that mobile B is within the area covered by unit 3.
- Switch unit 1 then sets up a call path between itself and switch unit 3.
- The mobile-to-base-station connections and the base-to-station-to-fixed-network connections are connected by the mobile switching units and the call can go ahead.

Notice that a significant amount of traffic passes through the fixed network as part of the setting up of the call. Indeed, even when calls are not in progress, there is still traffic passing through, as the base-stations are continually monitoring the location of the handsets that are switched on, and sending updates on the location to the databases.

In general, although this 'routing data' passes through the fixed network, it is not used by the latter for any form of call routing or control. The call control is managed by the mobile switching units, which usually belong to a different service provider than that of the fixed network. The mobile switches appear to the fixed network to be some variant of a conventional private switch. Thus the two networks are not really integrated in a strong sense. They do not share network service features such as network answering machines, etc.

What is a significant departure from simple telephony, however, is the volume of routing information passing over the network. Figures on this vary, and are often kept confidential, but the ratio of call-control information to that of the voice data is certainly at least 10%. This is an order of magnitude greater than for fixed handsets. The 'intelligence' of the mobile network has therefore to be similarly much greater than that of past networks. Mobile networks are the harbingers of the 'intelligent network' concept we discussed in Chapter 2.

TOWARDS UNIVERSAL PERSONAL MOBILITY

Perhaps the most significant architectural feature of the mobile network described above is not that it allows the terminals to be mobile but that it separates the identity of the user from that of a fixed-point telephone. We carry our 'phone number' around with us and we are allowed to access any terminal on the network, from any place. Our phone number becomes an access password and a means of configuring our current terminal to our personal requirements. Telephony begins to adopt the log-on concept familiar to computer users.

The beginnings of 'universal personal mobility' are emerging in a number of forms. Since the early 1990s, digital mobile telephony has begun to replace the earlier analogue systems. In particular, the 'Global System for Mobile Communications' ('GSM') has been adopted by nearly 200 operators in over 80 countries, although other standards are used, particularly in the USA and Japan.

GSM specifies allowable frequency bands (originally around 900 MHz and subsequently also around 1.8 GHz), channel bandwidths and modulation techniques for land- and ship-based services that, as far as possible, provide facilities (e.g. calling line identity, reverse charging, conference calls, etc.) offered by fixed networks without requiring the latter to be significantly modified. The initial speech coding technique encoded the speech at around 16 Kbit/s (13 Kbit/s in practice), although lower bit-rate systems have been developed. The standard is actively evolving: interoperability with the US Personal Communications service (PCS), interworking with satellite systems, lower bit-rate speech encoding and higher bit-rate data are all soon to be released, as is the facility to aggregate together a number of channels to provide even higher data-rates.

What is also emerging is the concept of compatible radio access across a wide range of scales (Figure 4.4).

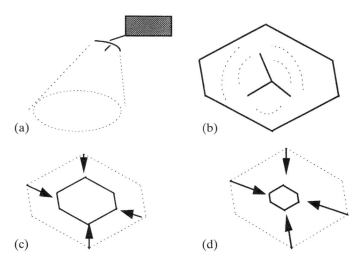

Figure 4.4 Cell-size strategics: (a) satellite cells for sparsely populated areas, cell radius 1000 km; (b) microcells (30 km radius) for low-population areas, with cells further divided using directional aerials; (c) cells shrink to less than 1 km for urban areas; (d) cells shrink further into picocells for use within buildings

The same handset will be capable of communicating via each and every one of these options. Expressed in this way, this is the viewpoint of the cellular network operators. Somewhat different approaches for the same end are adopted by two other operators: the fixed (i.e. wired) network providers and the new satellite companies.

RADIO IN THE LOCAL NETWORK

Fixed-network operators have a great deal invested in their fibre and copper networks, as well as a variety of mature network services. They want to capitalise on these but realise that the provision of new connections over the last kilometre or so may sometimes be dearer and slower to provide than radio alternatives. At least two options are available to them (Figure 4.5).

Where a large business is involved, requiring a number of channels, a fixed link, which assigns a wideband radio path allowing frequency division or time division multiplexing, can be provided (right-hand side of Figure 4.5). This is more or less 'transparent' to the users and appears to all intents and purposes just like a standard set of connections to the fixed network.

Where the sites are small, including domestic premises, a 'point-to-multipoint' system is used (left-hand side of Figure 4.5). Typically, this will offer a common radio channel in the 3.4–3.5 GHz band to a few hundred customers, providing privacy because each customer's channel of 32–64 Kbit/s voice or data is digitally encrypted.

Another approach to universal access is to consider integration of 'cordless telephony' – the short-range service provided between mains-powered base-stations and battery-powered cordless telephones within a single building (Figure 4.6). Within the building, the cordless handset continues to connect its calls directly into the fixed network, but it can also be used on the move, in which case it will connect via a local

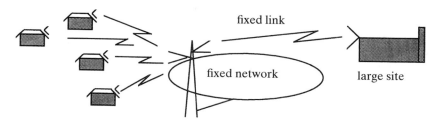

Figure 4.5 Radio fixed link and point-to-multipoint

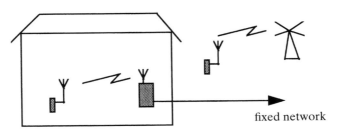

Figure 4.6 Integration of 'cordless telephony' and wireless mobility

area base-station, a cellular network, or even, as we shall see later, a satellite network.

As can be appreciated, these various technical options are part of the commercial competition between telecommunications companies, and their precise configurations are determined by commercial and regulatory regimes, which differ in different parts of the world. For instance, in the USA, one variant of the radio option is 'wireless fiber': 'long-haul' (trunk or inter-exchange) telecommunications companies (IXCs) are by law separated from 'local exchange carriers' (LECs), who own the wired connection between the local exchange and the customer. There is no such legislation regarding connections made by radio and, therefore, IXCs are very active in this area. Radio operating at 38 GHz has a very narrow width of beam, thus making it possible to run a number of systems, even from different companies, close to each other, to the extent that channels can radiate out from a central transmitter, separated by an angle of less than a single degree, without interfering. Rain is a slight problem as it absorbs radio signals at these frequencies, but transmitters can be designed to increase power automatically when the receiver sends back a 'low signal' message.

RADIO FOR MULTIMEDIA

So far, we have discussed the application of radio for providing narrowband voice and data services. A further use of radio is as an inexpensive way of providing wideband, multimedia signals to the home. Of course, we already do this extensively, as broadcast television. However, imminent changes in television transmission techniques and in narrowcasting networks offer new opportunities to radio and potentially new services to domestic customers.

Up till now, TV signals have been analogue, delivered either by terrestrial UHF transmitters, by coaxial cable, or by geostationary satellite. Only recently has a market been established for digital delivery, by each of these three methods. Digital television uses data compression techniques to reduce a standard quality TV picture and sound to a data-rate around 2 Mbit/s. This will mean that existing terrestrial and satellite systems (and some cable ones) will be able to increase greatly the number of channels they deliver (to several hundred). Significantly, all of the delivery systems are likely to use the same coding standards and will be competing with each other by offering services unique to their particular form of delivery. For instance, terrestrial and satellite can provide cheap and extensive coverage without digging up roads or requiring relatively expensive individual subscriber line equipment at the central distribution point. Cable and multimedia telecommunications companies, on the other hand, will probably emphasise the 'personalised' aspect of their offerings (true, no waiting, 'movies on demand', or fast, two-way interactivity, for example). One significant way they can also reduce their cost disadvantage may be through use of broadband (2 Mbit/s or higher) radio delivery to individual customers. The links could be broad in both directions, if high-quality two-way video were required (e.g. in a teleworking scenario), or broadband from the centre out, carrying one or more TV, movies-on-demand, home shopping, etc, channels, with narrowband return channels for interaction with these services.

DIGITAL TV BROADCASTING

One ubiquitous method of broadband delivery is already with us: TV. In its simplest form as a broadcast service it has, of course, the property that there is no selective transmission to individual households and there is no backward channel. There is also the problem that current transmission techniques, whether from terrestrial aerials or from satellites, make high demands on the rather restricted radio spectrum available for public broadcasting. Recent developments, however, bid to change this problem to some degree, by taking advantage of digital processing of the video and audio signals.

Today's TV broadcasting is 'analogue' (see Appendix, Chapter 11, for an explanation of this and other signal processing concepts). The basic minimum bandwidth requirement for transmitting an acceptable quality is around 5–6 MHz. However, it is necessary to shift the signal up in frequency and to modulate it in frequency modulated form (see frame, Chapter 3). This latter condition means that the required transmission bandwidth is, in fact, greater, around 8 MHz for each programme channel.

An alternative method, that has become feasible because of the ability to produce low-cost decoders for use in TV sets or associated set-top boxes (discussed in Chapter 5), is to code the signal digitally. It is not simply an issue of sampling the analogue signal and transmitting it as a digital bit stream – to do this would make the bandwidth problem worse. Instead, before transmission, the images are digitised and then subjected to 'digital compression'. This takes advantage of the fact that images are, by and large, not completely random, in two ways: firstly, adjacent points in any image often have similar colours and brightnesses; secondly, it is very rare that every part of the field of view in one TV frame is entirely different from what happened in the previous frame, 1/25 of a second earlier. This predictability means that we can get away with a lower data-rate than expected. (Again, see the Appendix, Chapter 11, for a more detailed discussion.) The long and short of it is that good-quality video and sound can be digitally encoded at a rate of a few megabits per second. Standard TV quality can be achieved at 6 Mbit/s, VHS video recorder quality at 2 Mbit/s and even High Definition TV at 24 Mbit/s. A set of international standards, collectively known as the 'MPEG' standards, have been agreed and coders and decoders working to these are now available.

This means that, typically, each one of the current analogue TV channels could be replaced by at least five digital TV channels. This in itself opens up the prospect of more minority viewing and 'staggercasting', that is, the repeated transmission of the same programming material on different channels, each channel starting a little later than the previous one, so that viewers can, to an extent, select their time of viewing, rather similar to 'video on demand' as described in Chapter 2.

Of course, the digital transmissions need not be restricted to TV programmes: 2 Mbit/s (or even somewhat less) is a very attractive rate for high-quality Teletext services, for example for home shopping or access to the Internet. As will be discussed in Chapter 5, the digital TV receiver or set-top box can also be connected to the telephone network, to allow for a backward channel from the viewer. This is the concept of 'Interactive TV'.

SATELLITE NETWORKS

As mentioned earlier, satellite broadcasting has become a key component in the provision of entertainment services directly into our homes. On the other hand, satellite communications networks have been, traditionally, competitors of submarine cables: a way of providing intercontinental trunk circuits, rather than for personal access. They carry bulk traffic to (mainly) fixed terminals – large satellite ground stations. These ground stations have been large partly because of the power and directionality required to communicate with satellites 36 000 km above the Earth's surface and, except for ground stations on the equator, at an oblique angle.

But just as very large scale integrated circuits and digital computer control have freed the telephone from its fixed tether, the same technology has made it possible for us to utilise satellites at orbits which are much lower and pass over higher latitudes. The consequences are revolutionary: the possibility of providing high-quality, wireless telephony from virtually anywhere in the world to anywhere else, the ability to provide low-delay paths that remove many of the problems of computer-to-computer communication that currently bedevil satellite links, and the potential to provide all of this at prices comparable with fixed-line solutions.

BASIC PRINCIPLES

The astronomical description of satellites puts them broadly into two classes: 'geostationary' and 'medium or low altitude'. All practical commercial communications satellites until recently fell into the former category. The term 'geostationary' implies that to someone on Earth, the satellites are always above the same spot on the Earth's surface. To achieve this, the satellite orbit is heavily constrained in two ways: firstly, it must be at a fixed height of approximately 36 000 km, in order that its rotational velocity is the same as that of the Earth; secondly, it can only match this velocity at all points in its orbit, if it travels around the equator.

The equatorial orbit means that points on the Earth at high latitude need much more powerful transmitters and more sensitive and lower internal noise receivers, and have problems with lines-of-sight because the satellite appears low down on the horizon.

However, perhaps the major problem is that of path-induced delay (or 'latency'): given that the speed of radio waves is 300 000 km/s, it takes of the order of a quarter of a second for a signal to go from sender to receiver, and the same time to reply. We are all aware of the adverse effects of this delay on telephone calls or intercontinental interviews on television; there is no doubt that it reduces spontaneity and also causes speakers to 'talk across' each other. What is less appreciated is its extremely adverse effect on the speed of data connections based on standard line protocols, such as TCP/IP, the cornerstone of Internet connections, as described in more detail in the frame. The consequence of combining Internet protocols with long

GEOSTATIONARY SATELLITES AND THE INTERNET

It is often assumed that increasing the transmission bit-rate is the answer to all of the current limitations on the performance of networked computers, but this is not true, at least in the case of Internet services based on the ubiquitous TCP/IP set of protocols that are used to establish and control the interconnection of the client computers and the servers from which they request information. A major consideration is network delay ('latency'), which can restrict the achievable data-rate to a small fraction of the available transmission rate. TCP breaks the data to be transmitted into blocks, called 'packets', and numbers and addresses them in such a way that each packet can be sent separately over the best available path at the time. At the receiving end, TCP re-sequences the packets in the correct order. The receiving computer is continually replying with 'acknowledgement packets', provided none of the packets has been lost or delayed too long. In the event that some have, the transmitting computer simply resends the data. In order to do this, it holds the data in a temporary buffer. It is normally efficient to have small buffers, typically 64 Kbits. This also means that only 64 Kbits can be in transit through the network for the error correction facility to operate. But the round-trip delay for a geostationary link is about 0.5 s. So, if a maximum of 64 Kbits are in the round-trip link at any one time, the maximum data rate is 64/0.5 = 128 Kbit/s, *irrespective of the bandwidth of the link.*

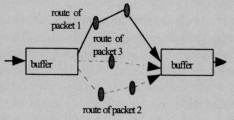

RADICAL OPTIONS FOR RADIO

latencies due to long path distance is that, typically, a maximum transmission rate of around 100 Kbit/s can be achieved even if the satellite link had infinite bandwidth. It is possible to modify TCP/IP and other latency-sensitive protocols, but there would be a problem with existing equipment. In any case, there is an alternative, with additional attractive properties: the low Earth orbit satellite, or 'LEO'.

LOW EARTH ORBIT SATELLITES (LEOs)

Low Earth orbit satellites circle the Earth a few hundred kilometres above its surface. At this distance, they are no longer geostationary and complete an orbit in a few hours. Thus, no satellite is continuously visible to any point on the Earth's surface, and connections between the ground and space are set up on a dynamic allocation basis to whichever satellite offers the best performance at the time. In many ways, LEO's turn mobile communications on its head: it is the base-stations that are continually moving, not necessarily the terminal. The satellites no longer adopt equatorial orbits; they orbit at a high angle to the equator, 'striping' across the Earth's surface, creating a cellular footprint across the globe, whose size depends on the radio frequency used and the geometry of the satellite aerial (Figure 4.7).

A number of radio connections, at a number of frequencies, are involved, depending on the service offered and the complexity of the satellite platforms. An advanced, complex example (Teledesic) is shown in Figure 4.8.

We can see that LEOs can communicate with 'earth stations' – these are major transmitting aerials which transmit and receive multiple

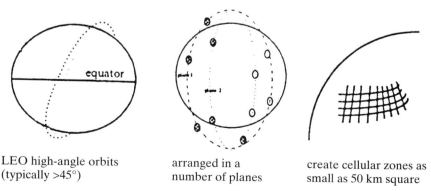

LEO high-angle orbits (typically >45°)   arranged in a number of planes   create cellular zones as small as 50 km square

Figure 4.7 The LEO 'footprint'

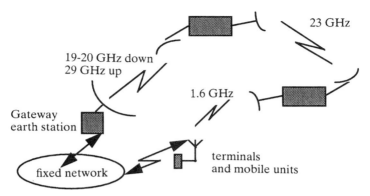

Figure 4.8 Satellite links

channels between ground and air and similarly have multiple connections to the fixed telephone network – and also with individual terminals which provide a small number of channels, perhaps just one. These terminals can have aerials on the roofs of shops or small businesses, or alternatively can be in vehicles, provide payphone points in remote areas or even be handheld portable telephones or computers. There are a number of alternatives for multiplexing multiple channels onto the up and downlinks: for instance, 30 voice channels can be combined together in one frequency band using time division multiplex or frequency division multiplex as described in Chapters 2 and 3. One interesting alternative for use particularly in mobile terminals is 'Code Division Multiple Access' (CDMA – see frame), which is an efficient coding method when we have a large number of terminals which, individually, are relatively infrequently used.

As shown in Figure 4.9, in the simplest case individual satellites merely act as radio relays for bridging large distances, essentially just extremely tall radio towers. They do not interact with each other, and all the routing between terminals is done by the earth stations, with overall control from one or more management centres. The earth stations can select to transmit to any of several satellites high enough above the horizon at any one time, according to traffic loading.

Later and more sophisticated systems are planned with switching available on board the satellites (Figure 4.10).

Satellites can now interact with each other, forming a highly connected grid of transmission paths from satellite to satellite, without involving the earth stations. This interconnected grid, it is claimed, will allow wideband, multimedia signals to be routed with very short end-to-end delay.

CODE DIVISION MULTIPLE ACCESS (CDMA)

Code division multiple access is a convenient way of combining a number of separate channels within one frequency band, without creating any significant interference between them. Unlike frequency division or time division multiplex which carry each signal in a separate frequency or time slot, CDMA signals from each source overlap in time and frequency, but in a way which leads to little interference.

Suppose we have one 'bit' of data. Usually we would transmit a simple signal to represent it (top diagram). But there is no reason why we could not transmit a more complex code to represent it. Three examples are shown here.

One's first thought would suggest that this was not a good idea. After all, the alternative codes

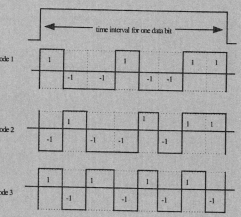

all change more rapidly than the simple representation, and will therefore consume more bandwidth. But suppose we multiply code 1 by code 2, segment by segment, and sum the answer:

$$1 \times (-1) + (-1) \times 1 + (-1) \times 1 + (-1) \times (-1) + (-1) \times 1 + (-1) \times (-1) + 1 \times 1 + 1 \times 1 = 0$$

Similarly, we find that codes 1 and 3, and codes 2 and 3, also yield 0 when combined in this way. However, two codes 1, or two codes 2, or two codes 3, each combine to yield 8. So, if receiving terminal n holds a template of code n within it and compares the incoming signal with its template by multiplication and addition (technically known as 'correlation'), it will produce a non-zero signal only when the received signal contains a bit destined for it. Thus, it is possible to add all the signals from all the active terminals together, and transmit them *on the same frequency band*, without any interference between channels.

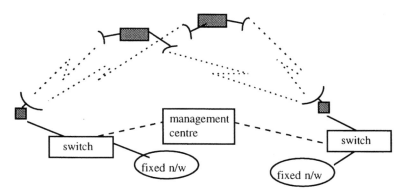

Figure 4.9 Routing via ground stations

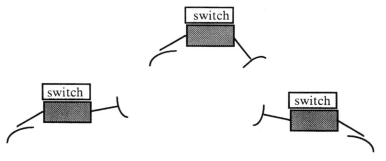

Figure 4.10 Active satellite routing

It has to be emphasised that satellite systems represent quite a high-risk solution to global communications, at least in terms of the traditional investment strategies of public telecommunications providers; indeed, most of the planned systems have no, or at most minority, capital provided by these sources. A number of doubts have been cast on the ability of satellite networks to cope with excess traffic volumes (e.g. in the case of national disasters such as a large earthquake), and on their ability to handle multicast transmissions. Nevertheless, the parties involved are major players in the world and undoubtedly very serious in their aspirations. Launching has already begun.

SOME EXAMPLES

At the time of writing, the market forecasts for LEO services, the make-up of the consortia and the solutions they offer are changing in a very volatile manner. The six frames shown here represent by no means an exhaustive list, and it is possible that some of these systems may never actually 'fly'.

ICO

This is strictly a 'medium Earth orbit' satellite, at 10 355 km. ICO plans 10 satellites in service, with two in view at any time. Satellites are connected to an earth station, which does all the switching, via TDM links. The earth stations are connected to the fixed, public switched telephone network. A narrowband (voice and data) service will be offered. Terminals (and much of the other technology) will be based on GSM standards; they will be capable of transmitting and receiving from terrestrial cellular base-stations and, in remote areas where these are not available, direct to the satellites.

GLOBALSTAR

Will have 48 satellites and 100–200 earth stations, each of which will connect to existing fixed and mobile networks. A narrowband (voice and data) service will be offered. Switching is under control of the earth stations. Code division multiplex is used for the mobile link.

IRIDIUM

Planned for operation in 1998, satellite launching has begun, with an ultimate total of 66 and around 2000 radio beams. It is a narrowband (voice and data) service. Although controlled by earth stations, signals can be switched via inter-satellite links, between each satellite and its four neighbours, in order to balance the traffic and optimise throughput. Mobile terminals can access the satellites directly, or via terrestrial mobile base stations. The architecture is heavily based on GSM and the principal communication will be between mobile handsets and the satellites or with GSM terrestrial base stations. Satellites and terminals communicate in the 1616–1626.5 MHz band, and 29.1–29.3 GHz is used for satellite-earth station connections.

SKYBRIDGE

This differs from the previous LEO examples in that it is intended to provide a broadband service. A constellation of 64 small satellites is used, linked to local earth stations that manage all the traffic within a radius of 350 km. It targets the market for Internet and Intranet (intra-company networks based on Internet technology), for multimedia online services for entertainment and professional use.

TELEDESIC

This is an extremely ambitious system, likely to cost around $9 billion and intended to offer optical-fibre quality (low end-to-end delay and extremely low error rates), bandwidth-on-demand from 16 Kbit/s to 2 Mbit/s. It plans to deploy at least 840 satellites, equivalent to a total capacity of 1 000 000 channels of 2 Mbit/s each. Although it will support narrowband voice and data mobiles, it will also offer wideband services at up to 2 Mbit/s for each terminal and rates of up to 1.2 Gbit/s to earth station gateways to public and private networks.

Routing of traffic is based on packet-switching principles, with ATM-based switches on board each satellite. Each satellite is a node in the packet network and is linked to two satellites ahead of it and two behind it, in the same plane, and with one each in the two adjacent planes. This creates a highly connected mesh, which can be reconnected 'on the fly' and is therefore resistant both to faults and to network congestion. Satellites hold incoming packets of data in a short buffer, until they have free capacity to transmit them onwards. The lengths of the buffers are also transferred from satellite to satellite across the network, and algorithms on each satellite work out the best of the eight possible directions to transmit the data, in order to minimise the end-to-end delay.

LEO – POLITICS AND MARKETS

The early success or otherwise of LEOs is deeply bound up with politics and cost: politics because, at least in theory, they offer a global bypass alternative to national fixed networks, and could be

> **CELESTRI**
>
> Celestri appears to be a pre-emptive response to Teledesic's broadband, 'multimedia' capability, by combining narrowband LEO design with the low-cost broadband and broadcast/multicast ability of geostationary satellites (GEOs). The aim is to allow regional (e.g. state or national) broadcasting with real-time interactivity, for services such as home shopping channels. The system will also allow the rapid construction of affordable, wideband computer-to-computer links for global businesses.

threatened by trade barriers initiated by governments at the insistence of the national carriers. In some countries, for instance the UK, it has been policy to encourage the development of terrestrial radio systems as a way of managing the introduction of competition. Satellite operators might destroy the balance of this plan. It is not surprising, therefore, that the LEO operators emphasise that they are intent on a policy of cooperation and complementarity with the established fixed-network and radio operators. This argument is more convincing in the case of collaboration with fixed-network providers; there is little doubt that, if LEOs are technically successful at the right price, they will end up competing with some of the terrestrial cellular operators.

5
Accessing the Network

TERMINALS FOR NETWORK ACCESS

The power and intelligence of future networks will not be realised unless we can access them by means of equally powerful and intelligent terminals, which are cost-effective for their purpose, and appropriate to their environment, whether static or on the move, for business or personal use. We are likely to see significant evolution in existing terminals as well as the gradual emergence of highly intelligent interfaces. We shall look at the intelligence that underlies these interfaces in a later chapter; here we shall examine the options for terminal architectures and some underlying technologies.

CONVERGENCE ISSUES IN THE DOMESTIC MARKET

Just as in the architecture of future networks there is a potential for convergence between the technologies and service concepts of telecommunications and computing, so we find the position with terminals for domestic use: only, in this case, there is tripartite convergence between telecommunications, computing and consumer electronics, with the emphasis on competition rather than cooperation.

The convergence is between three domestic products: the telephone, the TV set and the home computer (including the games computer). In this regard, it is interesting to observe that cultural studies of the telephone show that it is losing its 'body': if we view videotape footage of

Table 5.1 Software content of some consumer products

High-end TV set	600 Kbytes
VCR	256–512 Kbytes
Car radio	64–256 Kbytes
Audio set	64–256 Kbytes
GSM mobile telephone	512 Kbytes

today's films and TV programmes at scan speed, we quickly realise that, unlike a few decades ago, there are no freeze-frame shots of telephone instruments or people using the telephone. The frozen images are now of computers or people using them. In the 1990s, the screen is the 'icon' of communications, with the rival to the computer being the TV set. It is important to note that these two rivals do not differ much in technology, even less so with the arrival of digital TV transmissions. Even telephony is no longer primarily a hardware product, as Table 5.1 shows.

The distinctions are primarily cultural, with strong and persistent attitudes in the minds of users (see frame below).

What is perhaps just as significant is a shift in the communications potential of the TV set and the computer: originally, their characteristics and purposes were easy to distinguish from each other and from telephony. Figure 5.1 illustrates that TV sets were mainly for pleasure, even though the programmes might be instructional, and they were watched in a receive-only mode. Computer games were participative but played locally. Home computers were more 'serious', task-oriented devices, but rarely connected to the network. The telephone was the only really interactive device used over any distance.

The last decade has begun to alter that, with both the computer and the TV set developing aspects of remote interaction that challenge the role of the simple telephone, and we begin to see a possible migration path for domestic telecommunications away from the simple telephone, along three paths (Figure 5.2). In the sections that follow, we shall explore each of these paths.

ADVANCED TELEPHONY

The telephone we use today is a simple, reliable interface to a range of rather uncomplicated services that rely principally on voice and tone

> ## HACKER OR COUCH POTATO? – COMPUTING VERSUS INTERACTIVE TV
>
> Experts in cultural analysis have carried out comparisons of telephony, home computing and television, identifying a number of distinctive differences, which may persist even when all three products are connected to bidirectional networks:
>
> | | *TV* | *Computer* | *Telephone* |
> | --- | --- | --- | --- |
> | Interaction | public: watching together, discussing programmes with others | private: interaction is between yourself and a machine | private: a conversation between two people |
> | Mediation | heavily mediated: you get what the programmes determine | unmediated: you do what you want to; you access what you want | mediated: there are rules of conversation, a feeling that some behaviours are inappropriate |
> | Activity | entertainment, with some educational value, not heavily goal-driven | serious: goal-driven, work or serious hobby | social: heavily interactive |
> | Location | public rooms | study | private corner |
> | Appearance | consumer 'black goods' | technical | functional |
>
> If these conclusions are correct, it would indicate that no single, dominant, home platform will emerge. Instead, the reducing costs of the underlying hardware and software will lead to a convergence of technology but a divergence of products: the view-only TV set, the interactive TV set, the home-working computer, the computer for the kitchen that provides a recipe service and cooker control via a fat, flour and water-resistant interface, the garage car-diagnostic computer, the games machine, the portable communicator, the videophone and so on.

for their operation. It is likely that something approximating to the basic telephone will be around for several years, if only because it can provide a very low cost and highly reliable form of access. However, it is also likely to evolve into a number of distinctive forms, such as mobiles, multi-service and multimedia terminals.

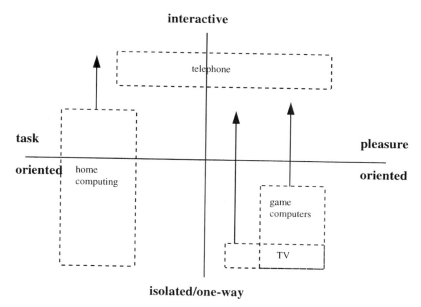

Figure 5.1 Platform convergence

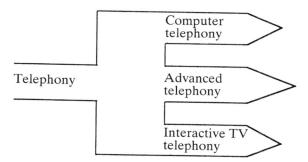

Figure 5.2 Evolution of domestic telecommunications

Mobile Telephones

Firstly, the telephone will become 'mobile'. The emerging standards for digital mobile telephony mean that the same handset that you use for wide area roaming could be used as a terminal within a local area or even within a building, thus providing a number of options, as shown in Figure 5.3. The portable radio telephone will be offered a connection to terrestrial and satellite links and, within a house, the 'cordless' telephones may communicate with a base-station within the building, that allows them to talk to each other as cordless extensions.

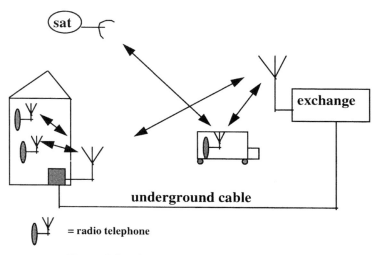

Figure 5.3 Options for mobile telephones

This base-station can also allow them to make or receive wide area calls because it is connected to the network, either by telephone cables or by a radio link. There are some quite powerful arguments for the move towards cordless telephony within domestic premises: firstly, as we have seen, the convenience of using the same instrument wherever you are, and secondly, although the cost of the mobile terminal will inevitably be somewhat higher than that of a fixed equivalent, this may be offset by the high cost of wiring. Unless buildings are comprehensively pre-wired to the required standard, an expensive refit of several hundred pounds is required. Many of today's householders are facing up to this fact as they look to install wiring for conventional telephone extensions; virtually all of today's houses are going to require a refit to take advantage of any wider-band services that will be developed in the next few years.

Multi-service Terminals

The advanced terminal will provide voice telephony, but will offer more. In particular, it will have a graphics interface, to make it easier to use the larger and more sophisticated range of facilities. For example, it is becoming rather difficult to set up shortcode dialling, retrieve messages from network-based answering machines, view whether you have diverted calls to other lines, understand the messages regarding caller identity services, and so on, using only recorded announcements and a simple keypad. Very soon we shall see significant marketing of advanced telephones with many of the features of personal organisers.

Apart from converting speech to and from electrical signals, telephones can provide a convenient, highly reliable, low-cost access point to the network for services other than voice. Many data services require modest bit-rates; one prime example is telebanking, and we shall see, in a later chapter, how electronic cash can be transferred over the network into smartcards that plug into a private telephone or public payphone.

Multimedia Terminals

Within a few years it will also be possible to offer moving video. For this to happen, we need to have available networks that can deliver at least a few megabits per second all the way to the terminal, at affordable prices. We have seen previously how copper, fibre and terrestrial and satellite systems are all competing to do this. As far as the terminal is concerned, there are still some issues of powering and display technology that require addressing before portable terminals can reach an acceptable standard. These will be addressed in the technology section at the end of this chapter.

MULTIMEDIA IN THE HOME: COMPUTERS AND TV SETS

The personal computer and the TV set are, technologically, very similar devices, although designed for different purposes. Originally, neither of them was equipped to handle two-way communication, but for some years now, networking of computers has been commonplace and TV sets are also being connected to telecommunications networks, through cable or other subscription channels. Currently, TV sets are evolving from analogue devices which contain special-purpose digital components into a much more digital form, with architectures that begin to resemble general purpose computers. Initially, much of this 'computerisation' is external to the basic TV set, in the form of a digital 'set-top box', but soon this functionality will be built into the set itself.

Set-top boxes (which, perversely, are often located *beneath* the set) are designed around general-purpose personal computers, minus, of course, the display and, usually, the keyboard (Figure 5.4).

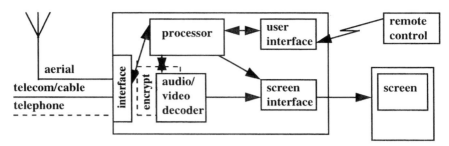

Figure 5.4 TV set-top box design

Signals are received, at megabit rates, from off-air transmitters, or over similar-speed telecom or cable circuits. There is also a telephony interface, currently for analogue telephony, but this could be digital. The digital signal is decoded by a special-purpose digital decoder chip, under control of the processor. There are a number of standard coding/decoding algorithms for digital television signals available, depending on the quality required. Notice also the dashed box labelled 'encrypt'. Although not necessary for the technical operation of the system, it is required in order to make money! The TV signal may be encrypted to make it unviewable without purchasing a key. One way to enable such a key is to allow someone to purchase it over the network. This is one possible use for the simple telephone interface. The picture and sound are then converted to analogue signals and passed to the TV set for delivery in the normal way. Users communicate with the set-top box using, typically, a hand-held infra-red remote controller.

The set-top box software structure also conforms to general-purpose computer principles set out in Table 5.2.

The software is 'layered' into a number of layers, each communicating principally with the layer immediately above or below it. Layers near the top are specific to the current application, whereas layers lower down are concerned with the operation of the hardware across a generic range of uses.

For example, in a home-shopping application, the user can browse through the 'content': a multimedia catalogue containing a number of still and moving images and sound together with product information and order forms. Apart from what is visible on the screen, the designer has generated a set of control information that links the various parts of the screen images; for example, selecting, by means of the remote control, a particular number that is shown on the screen to

Table 5.2 Set-top box software

Software layer	Residence
'Content', for example TV programme of film	Download on request
Control information specific to individual content, e.g. how to navigate through a home-shopping catalogue	Download on request
Generic environment software	Download during setting up of connection
General purpose, multimedia computer, operating system	Permanently resident
Hardware device drivers and network interface control	Permanently resident

represent 'next page', 'order form', etc., results in a command to the control software to retrieve and display the corresponding result. The content and the application control software are downloaded into the set-top box as and when required, during a session with the application.

At the very bottom of the layered structure are the network control software and a basic-level operating system; the latter directly controls the hardware elements of the computer (memory, input–output, the running of programs) and provides, to the next higher layer, a set of instructions that make it easier to control these hardware elements through symbolic instructions such as 'read', 'write' and so on.

There is also a layer between the operating system and the specific item of content: the 'generic environment'. The best way to define this is to say that it is the software that distinguishes the set-top box from a purely general-purpose computer and at the same time allows the box to take on a variety of functions or to be updated to take into account new functionality. For instance, suppose that some time after you bought your set-top box, you hear that a new service has been developed that can automatically learn your interests in fiction and negotiate with online libraries and bookshops to identify books that meet this profile. Because your set-top box has been designed with a standardised generic environment, you can download this application into the box, using very simple, foolproof commands, and thereafter run it automatically. The software may even be intelligent enough to update itself with the names of the participating libraries, by periodically downloading a directory.

This approach, whereby software is not always loaded into the equipment at the initial time of purchase but is instead swapped in and out and downloaded across the network, is particularly attractive for consumer goods, because it can be arranged so as to minimise the user's involvement in configuring and maintaining the system. The argument can be carried further: since in many cases there is a requirement for a domestic terminal that can very easily be configured to one of a range of special-purpose tasks – for shopping, access to cooking information, control of central heating, etc. – rather than a general-purpose computer, why not strip the terminal down to its bare essentials to reduce cost, and provide all the software maintenance and updating via the network?

THE NETWORK COMPUTER

Computer terminals, particularly those incorporated into networks, have evolved from two, quite different, operational principles. One is based on the idea that the terminal is very much a simple device, whose chief function is to buffer and otherwise condition the data that is chiefly processed by a much larger computer that acts as the hub to the whole network. This we can call the 'centralised model'. The other principle, which we can call the 'personal computer model', assumes that our computing requirements began with a set of computer users who processed data individually on their own machines, which they had purchased out of their own departmental or personal budgets. Subsequently, they felt a need to link their machines together, either for intercommunication purposes or to share expensive resources such as printers or large file-stores. Both of these approaches have advantages and disadvantages, some of which are listed in Table 5.3.

Over the last few years, the balance of the argument has lain with the personal computer: it has freed users from centralised control and budgeting; in most companies, the PC has given the functional departments (sales and marketing, finance, etc.) the opportunity to move forwards with automation at a pace that would have been impossible had everything been centralised in an IT function; it has also opened up the domestic market for truly personal, even recreational, computing. But there are also obvious problems: the increasing pace of change of software updates has led to many compatibility problems that can only be resolved by buying expensive upgrades. There has also been an ever-increasing hardware specification to handle this

Table 5.3 Centralised and personal computing

Centralised model	Personal computer model
Control is centralised, with little room for flexibility	Users can largely choose the configuration and software components to meet personal needs
Systems are consistent and interworking is guaranteed	Systems differ widely and interworking can be a problem
Software updates are maintained across all terminals	No control on updates between users
Budgets are centrally planned and need long lead-times	Purchases are under local control
Hardware requirements are minimised (e.g. system memory costs)	Communications costs can be minimised by processing locally

software, which means that the base price for a domestic computer has remained resolutely around the £1000/$1000 mark, which may be too high to give saturation penetration of the domestic market. To address that problem, some computer companies are considering the desirability of a very basic machine, the 'network computer', most of whose software is downloaded over the network, perhaps every time the terminal is used. The arguments for and against this approach are very much clouded by market politics. An extreme view from the network computer camp is that this can lead to the virtual demise of the PC operating system: all that the network computer need hold is a very simple operating system kernel to handle the loading up of the system and a simple interpreter that can run downloaded programs.

This is probably a deliberate over-simplification. But implicit in all of these arguments is the assumption that it will be cost-effective to download a higher percentage of the operating code than in the past. This will probably be true if the transmission cost per bit goes down significantly, as it will. What will be even more significant is if the transmission rate also increases, particularly if this increase is dramatic. This will alter the balance between local and remote processing.

SEPARATION OF CONTENT FROM PROCESSING

As described earlier, this already happens, to some extent, with cable TV networks, but will soon become more evident with the development of wideband digital transmission and the digital set-top boxes and networked computers connected to them. The software in the local terminal is concerned with the control, decoding and presentation of multimedia data that is retrieved from a remote source. In the longer term, this data may not just be movie or home-shopping information; it may be virtual reality images, perhaps in three dimensions, that are generated on remote, powerful computers and networked out to users. The balance between local and network processing will be one of competitive reductions in cost of computer terminals versus transmission.

OPTIONS FOR DOWNLOADABLE CODE

The vision of the network computer camp is the possibility of providing a rich environment of downloaded computer code into a basic, low-cost machine. The short-term instantiation of this vision can only be based on current, telephone line bit-rates, that is, a few tens or perhaps hundreds of kilobits per second. In this case, we could probably imagine application software, such as a training package, a route planner or a simple graphics-based multi-user game. Also, we might be prepared to wait several minutes to download an update to a software package, perhaps even a new version of an operating environment.

However, plans for digital radio links, the downloading of software from digital satellite, and video on demand over copper telephone cables or cable TV networks, may soon offer bit-rates of the order of 2 Mbit/s, as described in earlier chapters. From the user's point of view, this facility would probably be perceived as fundamental shift if it allowed one to invoke the effects of the downloaded program within a second or so of its being requested. This may raise the more radical option of distributing the active, real-time computing operations across the computer in the home and, simultaneously, the central computer. Indeed, we might even consider an application running simultaneously across a number of computers that are tightly controlled via fast communication networks and protocols connected to special-purpose central computers. This is the most extreme vision held out by the 'centralists', although it is, as yet, by no means clear what applications would most benefit from this, in the domestic market.

THE 'MULTICOMPUTER HOUSEHOLD'

Perhaps a more likely scenario is the growth of the 'diverse multi-computer household', where the domestic PC continues to grow in power and memory in line with Moore's Law, thereby not significantly reducing in price, but is accompanied by a number of devices that are constructed according to consumer-product principles – a minimisation of the number of circuit boards, special-purpose large-scale integration, hardwiring rather than plug-in components and boards, no extension ports, etc. These devices will be network computers, designed within a market dominated by 'substitution pricing', where they compete, not just with other computers, but with other totally different things (holidays, clothes, whatever). The hardware may be more specialised – the kitchen or garage computers mentioned in the 'Hacker or couch potato' frame earlier in this chapter, for example – and their limited memory (a function of the need to contain cost) will require the ability to download software when required.

One advantage other than cost that this approach can give is the minimisation of the need for owners to carry out their own 'systems administration': most of the files and virtually all the executable software is held centrally. This means that users do not have to bother about looking after sets of floppy disks, for example, or need to be conversant with how to upgrade their system to the latest version. The latter is an important point when a number of users need to share common resources. One example might be a teleworking scenario, involving non-IT personnel, who were simply required to use computers for some repetitive operation. The application software and the necessary data could be downloaded into any networked PC in the knowledge that the operation would be consistently carried out. Another market that has been identified for network PCs is in education, where costs can be minimised by using the cheap machine, and training packages can all be guaranteed to be consistent and under version control because they are retrieved from the central repository.

TERMINALS AND CONNECTIONS FOR BUSINESS COMMUNICATION

We said above that the future for domestic terminals was driven by communications, computing and consumer electronics. Consumer electronics is unlikely to play a major part in the business area, and, as implied in the scenario at the beginning of this chapter, there will not be as much obvious change in the appearance of business terminals, but the impact of communications technology on the role of business computing will undoubtedly be dramatic within the next decade, just as the requirements and capabilities of computing will be to the networks. The workstation or personal computer will remain the workhorse of the business person for many years yet, but the technology it contains, its power and its mode of connection to the local and wide area networks will be profoundly different from those of today.

HOW THE COMPUTER IS AFFECTING OFFICE TELEPHONY

First, however, we should look at the effect of the computer on internal telecommunications. In the past, internal telephone networks, controlled by a human operator or by a 'private automatic branch exchange' (PABX), were the sole means of rapid communications over a distance within the company, or with external customers who were not met face-to-face. Telephone extension numbers were listed in a phone directory and the PABX was configured by the company's telecom engineer to route calls to designated extensions. Until 20 years ago, this was mainly done by hardwired connections; since then, software programming of the PABX has become much more common, but PABXs even today tend to have a special, limited functionality. When a customer did get through to an enquiry agent, the latter was often supported only by paper files of instructions, or, more recently, by having access to a computer terminal on a local area network which was totally independent of the telephone and PABX.

In recent years there has been the recognition that the two networks – telephony and computer LAN – can be integrated together to provide useful synergy. To give two very simple examples: it creates a much better impression if the enquiry agent has instant access to the customer's files and knows what purchases the customer has made, the current balance in the account and so on. The need to ask the customer for a name and then key it into a terminal can be done away with, if the caller is equipped with calling line identity and the PABX can extract this information, passing it to the system computer even before the call is answered. The agent who answers the call can therefore be already primed with all the details necessary. The time of the call can also be automatically entered into the computer, for example to maintain fault records or for scheduling a visit.

Another example involves 'out-bound calling', in a telesales operation. Computer analysis of sales and marketing data can lead to the creation of lists of prospects' telephone numbers, a short-list of products that each might want to buy, and a third list of sales agents, their skills, their rotas and their extension numbers. Cooperation between the PABX and the computer can lead to automatic call generation, distributed to the correct agent extensions together with a suitable script on the screen that they can use when the prospect answers the telephone call.

These are examples of 'computer/telephony integration', commonly known by its initials, CTI. Instead of building PABXs with inbuilt, specialised intelligence, the PABX is designed to consist of a simple, but reliable, switch for setting up and closing down calls, which is connected to a general-purpose computer which carries out all of the intelligent call-control as well as its normal computing functions. Because the computer is a standard product, its hardware and software are considerably cheaper to develop than that of a specialised PABX and much easier to integrate with other application packages.

CTI can be achieved in a number of ways. At the bottom end, a small system can be constructed using a personal computer and a single telephone, as shown in Figure 5.5.

Two configurations are shown: in one, the telephone is used to interface to the network; in the other, the computer is the dominant unit, with telephony relegated to handset functions. The choice of configuration is partly dependent on the desire to minimise the

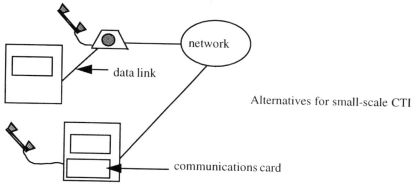

Alternatives for small-scale CTI

Figure 5.5 CTI – a small installation

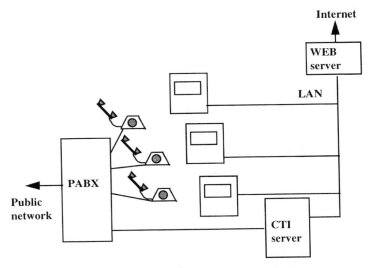

Figure 5.6 CTI – a large-scale installation

disruption to any existing equipment, but it is also 'cultural', depending on the attitudes to telephony and computing within the organisation.

On a larger scale, CTI involves combining traditional PABX functions and hardware with traditional computer networks (Figure 5.6).

In the figure, there is no direct physical connection between the telephones and the workstations; instead, functional integration is achieved by connecting the corporate LAN to the PABX via a 'CTI server', which can communicate switching requests from the computers to the PABX and call data from the PABX to the computers via a set of communications protocols which can be proprietary but are increasingly becoming standardised.

The intelligence embedded in CTI private networks such as these is considerable, and allows independent organisations to construct a number of features which, only a few years ago, would have been considered the domain of the 'intelligent networks' offered by public telephone companies (see Chapter 2). Moreover, shown in Figure 5.6 *but not usually shown when the diagram is drawn by telecom engineers*, is the gateway to the alternative to the telephone network – the Internet. Precisely what can be achieved additionally through this connection is still, to an extent, speculative although in Chapter 2 we have alluded to possibilities for Internet telephony or least-cost routing.

ADVANCES AND REQUIREMENTS IN TERMINALS

For the next decade at least, it is likely that the most common device for business computing will be something that is still recognisable as having evolved from the personal computer or workstation that we know today. From a product point of view, computer terminals are not particularly complex devices. Their major building blocks are shown in Figure 5.7.

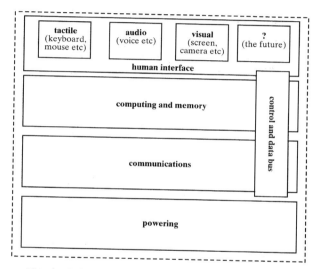

Figure 5.7 Basic components of a terminal

POWER SUPPLIES

At first thought, power supplies are only an issue with portable equipment for use on the move, or at least away from the office. However, this is not always the case. If, for example, you wish to achieve maximum flexibility in office or store layout, you need to be untethered from the mains supply; the benefits of radio or optical LANs – that they do not need to be plugged in – are somewhat diminished if you still need to provide trunking for power.

Today most portable electronic equipment is relatively power-hungry: GSM telephones consume about 1 W at peak. Laptop computers operating fully consume power at a rate of 15–20 W and the batteries last only 2–3 hours without recharging. Small disk drives also consume power at the level of a few watts, and portable computers have smart circuitry which shuts down the drive except for brief periods of access. Miniature disk drives are available but all-solid-state memory is also a goal for portability, the biggest issues being the performance/cost ratio and access time. Most integrated circuits consume significant power only when changing state; fortunately, progress in reducing gate dimensions, and hence power consumed, has been consistently successful and this is likely to continue. However, the trend to higher and higher data-rates will operate against this. Finally, higher performance displays will set challenging targets for power sources.

Solar energy is a questionable solution: a 30 × 30 cm array of single crystal silicon would be just sufficient to power the 15 W laptop mentioned above. Higher efficiencies can be obtaining using III–V semiconductors but the cost/performance ratio becomes even less favourable. The efficiencies of cheap solar cells made from amorphous silicon are typically only half those of their single crystal counterparts. In the longer term, solar conversion using synthetic organic molecules which can be coated onto a roll of material (artificial photosynthesis) may lead to lower-cost solar energy production.

We probably have to dismiss other ingenious suggestions such as *thermoelectricity* from the human body, which can give only a few milliwatts, or miniature electricity generators which convert body motion to electricity (microwatts). A more concentrated form of energy would be by direct chemical conversion, for instance by oxidising butane gas catalytically to generate electrical power, but this requires high temperatures and would not be easy to use in a pocket terminal, for example!

The conventional predictions for portable power centre on relatively few technologies. *Primary batteries* produce power by an electrochemical reaction which eventually destroys the chemicals involved, irreversibly. Newer types of primary cell, such as zinc–air, have up to 10 times the capacity of the established mass-market cells but are not yet in widespread use.

Secondary batteries can be recharged many times by passing a current through them in the reverse direction, thus partly reversing the chemical reaction. Gradual deterioration limits this to about 1000 charging cycles. The NiCd battery is being superseded by nickel metal hydride, which contains no toxic metals. Lithium ion polymer batteries are now emerging as viable products. These have a much greater power to weight ratio and energy density and can in principle be moulded to any desired shape. These will form the basis for power for portable equipment in the next few years.

The main difference between a *fuel cell* and a battery is that the fuel (e.g. hydrogen) and oxidant (e.g. oxygen) are not integral parts of the cell but are supplied as needed to provide power to an external load, while the waste products are continuously removed. The core of a fuel cell typically works at a temperature of 100–200°C or more. Fuel cell technology has made great strides in recent years but it is aimed mainly at power in vehicles.

Although of minor importance in peacetime in first-world countries, *clockwork power* cannot be dismissed out of hand, when we consider third-world countries or battlefield equipment. Trevor Baylis, the inventor of clockwork power for radio, has demonstrated that it can also be used to power laptop computers, and the US military are considering it for tactical use. As with solar power, the limited capacity and (in the case of solar power, at least) the relatively high cost per watt, are somewhat cancelled out by the low maintenance, a critical factor in backward regions or when under fire.

HUMAN INTERFACES

Similarly, there are no really radically new interfaces observable today that are likely to replace the screen, the keyboard and the mouse, or variants thereof, within the next decade, with the possible exception of voice recognition in environments where it would not be intrusive. The major changes are more likely to be in the use that is made of them, particularly the visual display screen. Screens provide a good medium for rich and intuitive interaction between people and computers; therefore, owing to the emergence of new technologies, they are likely to become more prevalent and applied to an increasing range of communication equipment, including quite basic telephones, and also larger in size.

A number of different technologies are described in Table 5.4 and one prediction is made in Table 5.5.

INTERACTIVE VIDEO

One increasingly common interface will be the TV camera. Already this is in regular use by a number of international companies, for

Table 5.4 Future display technologies

- *Cathode Ray Tubes* Dominant for almost 100 years, being simple and cheap to produce. Bulky and power-hungry, usually giving a picture of only moderate brightness and size.

- *Field Emission Displays* Like flat electron tubes with millions of microtip electron emitters on the back plane (hundreds per pixel) and a phosphor-coated front screen. A few millimetres thick and consume only a few watts. Demonstrators with 100 to 150 mm diagonal have been produced and commercial exploitation is beginning. May emerge as mass-market products in the next decade.

- *Liquid Crystal Displays* Use various organic molecules whose optical properties change in an electric field. The dominant medium for portable devices: compact, low power and drive voltage. Difficult to achieve good yields of large, full-colour LCDs with fast refresh rates.

- *Light Emitting Diodes* Red, yellow and green established and recently good progress in making blue LEDs using III-V nitride semiconductors. Currently used for giant screen displays with one LED per pixel. An exciting development in LED displays is light emitting semiconducting polymers (LEPs), tuned to red, green and blue. External efficiencies are currently only ~2%, and the material degrades with time but may in time enable the development of low cost, flexible displays with low power consumption.

- *Electroluminescent Displays (Els)* Generally useful only for low information content displays, such as in instrumentation. Mostly monochromatic, although red, orange, yellow and green emitters can be made relatively easily and developments may see high information content Els. Compact, sizeable, good brightness, high writing speed and grey-scale capability. High voltages required and power consumption is high. Producing full-colour displays over large areas reliably and cheaply will remain elusive unless novel technology for continuous deposition and device patterning can be developed on sheet material.

- *Vacuum Fluorescent Displays* Use electron bombardment onto very thin layers of highly efficient phosphors, coated directly onto transparent anodes. Operate at lower voltages than other emissive displays and have good brightness. However, they are suitable only for low information content displays, since the cost escalates for large area and high pixel counts.

- *Plasma Panels* Employ the glow discharge that occurs when ionised gas undergoes recombination and promise to fulfil mass-market solutions for large flat displays during the next decade. Colour displays 1 m in size and of high quality have been demonstrated and, while relatively inefficient

Table 5.4 *Continued*

and hot, give bright coloured images with good viewability. New hybrid devices, in which the plasma cells are used to switch an LCD with backlight illumination, are being developed to overcome the difficulty of fabricating a fault-free active matrix semiconductor backplane for LCDs, by replacing the transistors by plasma cells driven by simpler grids of metal electrodes.

- *Nanotechnology in Displays* Light emission is extremely inefficient in bulk silicon, but more efficient green to red luminescence can be achieved in silicon microstructures, such as porous silicon, created by an electrochemical reaction on single crystal silicon, producing arrays of silicon pillars, only nanometres wide, in a coral-like matrix, in which motion of the electrons is confined. A number of challenges remain, including stability and lifetime, and the production of blue emitters. A similar principle is involved in optical emission from nanometre-sized particles of semiconductors (quantum dots).

Table 5.5 Display technology

Impact in 0–5 years	*Impact in 5–10 years*	*Impact in >10 years*
• LCDs competing with CRTs	• Large plasma panels	• Polymer EL displays
• Medium-sized plasma panels	• Medium-sized field emission displays	• Photonic nanostructure displays
• LCD wearable displays	• Simple polymer LED colour displays	

Source: BT Labs

video-conferencing or for remote inspection of equipment or processes. Video-conferencing of good quality, particularly if it involves more than one person at each end, requires bit-rates of the order of 2 Mbit/s. It also requires circuits with low end-to-end delay. Usually the conferencing equipment is set up in a special studio, or at least involves a fairly expensive, special-purpose coding and decoding unit ('codec'). Where there are more than two parties to the conference, 'multipoint conference bridges' are required, sometimes supplied within the network or, alternatively, when a company is a sufficiently frequent user, its own equipment, held on one of its sites.

Today, it is also possible to buy desktop video equipment that runs on a PC. This provides 'adequate' rather than good picture quality, when delivered over basic rate ISDN telephone circuits at 64 or 128 Kbit/s, and usually somewhat poorer quality when delivered via the Internet. (Sometimes, in the last case, it does not work at all.) The better quality desktop equipment comprises a mixture of hardware and software; the processing speed of most desktop machines is not quite fast enough to carry out the video coding and decoding in software only, but, whatever the method employed, video transmission at sub-megabit per second rates always involves significant processing delays, such that the receiver views an image which is a couple of hundred milliseconds old. One obvious consequence is the need to delay the audio channel by the same amount; a less obvious but more problematic issue is the effect of this delay on the subjective acceptability of the received signal.

Conversations involving audio delayed by this amount can be slightly disturbing, with the problem that both parties, feeling the need to fill the vacuum of silence, begin talking at the same time. This effect is, of course, much magnified when added to the hundreds of milliseconds always introduced by circuits which travel via satellite links. The hesitant response introduced by these effects does not enhance the credibility of the distant party. This is further adversely affected by another artefact of desktop video – the location of the camera. This is usually placed on top of the monitor, as shown in Figure 5.8. The user normally looks directly at the monitor, to view the image of the other party to the conference. To the camera – and thus to the person viewing the image at the far end – this looks as if the user's eyes are directed downwards, avoiding direct contact, sending out body signals that imply shiftiness or unwillingness to communicate.

A number of solutions have been employed to get over this problem. Perhaps the simplest is to interpose a half-transparent mirror at 45° between the screen and the user (Figure 5.9). The image of the user

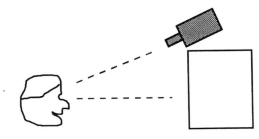

Figure 5.8 Conventional placing of video camera

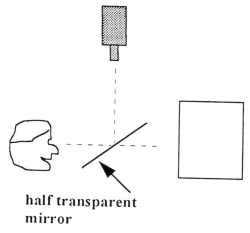

Figure 5.9 Camera placement for eye-to-eye viewing

is reflected by the mirror into the camera. (Of course, it needs to be reversed once more, either by using another mirror, or electronically, in order to give an unreversed final image.)

3-D VIDEO

A further way to improve the naturalness of transmitted video images is to use some form of 3-D. True 3-D, in the sense of high-quality, free-standing images that we can walk round and examine from all angles, is still beyond our grasp, and will, in any case, almost certainly await delivery of Gbit/s bit-rates. However, for viewing scenes of restricted size – particularly width a depth of more than a few metres – simpler, stereoscopic systems are quite feasible. These work on the principle of capturing the scene using two (or more) cameras from slightly different angles. Alternatively, if the images are not of real objects, for example in the case of the simulation of an engineering component or in a computer game, they can be generated from more than one 'virtual viewpoint'. What is now required is to present one of the viewpoints to the observer's left eye and the other image to the right. A number of ways of doing this are possible:

- Images can be created by the superposition of light projected by two projectors. The light from one projector passes through a filter that permits only vertically polarised light; the other projector has a horizontal polariser. To view the image, it is necessary to wear

spectacles with one eyepiece containing a vertical polarisation filter and the other horizontally polarised. The image quality can be good, but there is the inconvenience of finding enough space for projection and viewing.
- Another solution, which also requires the wearing of spectacles, is to adapt a monitor to run at twice normal speed and to display the right-hand and left-hand images one after another. This time, the spectacles are constructed from liquid crystal material which can be rendered opaque or transparent by the application of an electric field applied to transparent electrodes on either side of the glass. These electric 'shutters' are synchronised to the display monitor, to alternate open and shut, such that the left eye sees only the left image and the right eye the right.
- A third alternative is to use multiple cameras to capture the image and then project it to the viewer through a panel of lenticular lenses (Figure 5.10). If the lenses are lined up carefully with the images, then it is possible to allow the left eye to see only the left-hand image and the right eye the right. The greater the number of viewpoints required, the greater the number of pairs of camera positions.

ADVANCED ALTERNATIVES FOR NETWORK INTERACTION – TELEPRESENCE

The earlier sections have concentrated on the use of intelligent communications for the passing of *information* with little attempt to

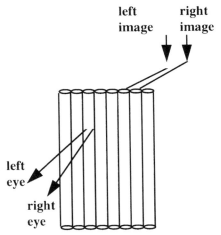

Figure 5.10 Lenticular lenses for 3-D imaging

immerse the users within the remote scene and no attempt to give them physical control over it. A range of tools, collectively describable under the name of 'telepresence', attempt to do just that. The aim is to use the network and its attachments in such a way that users are no longer at a distance from the other end and can interact directly with it, as if they were near. Essentially, we are creating remote senses and powers of motion. We must be careful that we do not get too carried away by these concepts: the invention of motor-powered factory machinery did not give rise to things with arms and legs that manipulated tools in exactly the same way as human beings. The answer to all remote action may not be the body-suit that we climb into; it may be something much more conventionally machine-like. Nor must we confuse telepresence with robotics; in many ways they are competing solutions: in a tele-action mode, we are using human pattern recognition and intelligence to control processes at the other end of a network; in robotics we are removing at least some of the need for the network, by providing artificial intelligence locally. In practice, it is likely that the two will cooperate, with the recognition and decision-making split between the machine 'at the coal-face' (and, indeed, robotic mining might be a good example) and the human safely located a long way away. In the brief overview that follows, we shall look at some of possibilities for remoting the various human senses. If future networks are truly to dissolve distance in all its aspects, we shall need also to touch, smell, and hear ambient and localisable sound, if we are to create a total multimedia experience. Beyond that, there are extra senses that we do not naturally possess, such as the ability of some animals to see in the infra-red spectrum, the ability of instruments to monitor blood-flow without invasive surgery, and others, that would go together to give us a super-human 'telepresence', the ability to interact with remote environments across the network.

SMELL

What may not be immediately obvious is why we might want to create an artificial sensory system that would simulate the human sense of smell, but a number of practical justifications have been put forward: the sensing of smoke or dangerous chemicals is one example; access control by recognising people's individual smells is another; tele-medicine provides another example in the case of certain illnesses which give a characteristic smell to the breath or skin, that we would want to pass to the consultant at the other end of the communications link.

The ability of the human nose to detect and classify smells is, as yet, not fully understood, but it is known to be highly sensitive and selective (although recent research has shown that evolution is leading to a reduction in this capability – perhaps a reason for finding electronic replacements). Artificial noses have not yet reached this level of performance but can still detect quantities as low as one molecule in 1 000 000 000 parts. They are constructed from a number of conducting polymers whose electrical resistance changes when exposed to the vapour. By changing the composition of the polymer, it can be made to react to vapours from different substances.

TOUCH

Virtual reality (VR) games are currently the biggest market for touch simulation technologies, and much of the effort to date has been driven by this market. There are a number of other feasible applications waiting in the wings, perhaps requiring better technology, or simply greater awareness and experience among their potential users.

We can look, for example, at its use in healthcare. Medical examinations can be carried out at a distance, at least in part, by means of video and dialogue with the patient, but there are times when a physical examination is necessary, to feel the hardness and mobility of a suspected tumour, to access abdominal swelling, to locate a baby's position in pregnancy, and so on.

A possibly beneficial side-effect comes about by interposing a layer of indirection between the user and the items to be touched: it is possible to change scale between the actual and perceived pressure. This might allow the user to carry out much more delicate and tiny manoeuvres than could be achieved in a 'hand-on' manner. In the ultimate, it allows people to feel their way around virtual worlds which do not actually exist. One application for this is in chemical engineering, where spatial models of complex molecules can be viewed and manipulated remotely at the same time. The strength of the molecular bonds is built into the model and the chemist can feel and see the effects of moving the components.

The simulators available today are, in the main, simple enhancements to data-gloves that measure the positions of the parts of the hand and can be programmed to oppose the motion by means of air pumped into pockets on the glove. Apart from lacking precision, they are rather expensive to produce and, together with their drive-equipment, are bulky.

A possible improvement on air pockets is to use a laminated glove that encloses a layer of electro-rheological fluid. Fluids of this type offer little resistance in the absence of an electric field, but stiffen significantly when one is applied. The increase in stiffness is related to the strength of the field and is reversible. At present, the requirement for high voltages make them unsuitable for most applications, but research is underway into new materials.

Of course, if we are to provide feedback to the sense of touch, we must also measure location and movement of the part that is doing the touching. Elements embedded in gloves are the most popular way today, but microwave radar has been suggested, as has optical recognition of gestures. Both of these require a certain amount of intelligent image processing, which is discussed in the next chapter.

If we are accurately to imitate the discrimination of our senses, we need to realise how precise they can be. For instance, the fingertips are capable of millimetric accuracy (Figure 5.11).

We can combine the measuring of movement and the consequent sensation of touch by using tiny mechanical systems. Microengineered components based on silicon substrates have been proposed. One such example is shown in Figure 5.12. The movement of the cantilever creates an electrical signal which can be used to sense the motion. In turn, the system can pass a current through the solenoid which generates the appropriate level of opposing force.

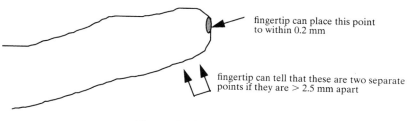

Figure 5.11   Sense of touch

SOUND

A major component of our everyday interaction with the real world is currently neglected in most virtual interactions: the role of realistic

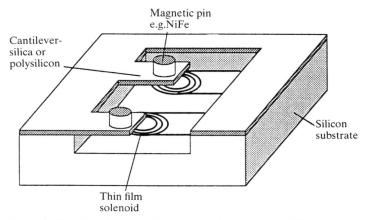

Figure 5.12 Touch sensation generator (source: BT Labs)

sound. Unlike reality, we receive sound at the workstation either as discrete bleeps from the direction of our computer or as telephone quality, located again at a point. The natural clues regarding spatial location or relative importance are removed. In providing a more natural acoustic ambience, we have, of course, to avoid disturbing the environment of others. This is one reason why today we constrain acoustic interaction in the business environment to that of earphones or telephone. However, consider the possibilities of realistic sound in the case of a virtual meeting some time in the future: with a wall-sized monitor we begin to separate out the participants in the meeting and it would greatly add to the naturalness of the proceedings if the spoken sounds appeared to be collocated with the images of the people that uttered them.

Another application is in creating realistic environments for collaborative working: experiments have demonstrated the benefits of creating 'logical spaces' that correspond to the imaginary desks and offices that would be populated by the collaborators were they to be collocated in reality. In this virtual office, it is helpful to provide each participant with an 'aura', an area of personal space, into which other people intrude only if they want to interact. Rather than create an all-or-nothing aura, it is more natural and effective to provide a sense of gradual encroachment. One way to do this is to increase the perceived loudness of speech as one person approaches another. In the virtual office, for example, two people can have a conversation in the background (as perceived by a third), but when they navigate their virtual shapes towards her, using a mouse, say, then the volume of the conversation rises.

One sought-after, though unlikely, scenario is the possibility of an office space equipped with a 'sound bubble', an area surrounding the user, in which users can hear realistic spatially located sounds without headphones, but with the sound field constrained within the immediate vicinity, so that it does not intrude on others. To do this effectively would probably require the design of very special cubicles and precise location of surround-sound equipment. Relying, as it would, on dynamically modelling the acoustic response of the room and its fittings, it would also be sensitive to any moving person or object.

THE NETWORK PC IN THE BUSINESS CONTEXT

Earlier in this chapter, we described a 'minimalist approach' to terminal design, based on keeping as much as possible of the software and hardware centrally. There we were referring specifically to the domestic scene, where the supposed advantages are in convenience of use and in minimising hardware costs. A somewhat different set of arguments can be made for such networked terminals in the business environment.

The benefits do not, perhaps surprisingly, come from reducing the cost of the terminal. It is unlikely that business machines will be able to make do with the reduced functionality that might satisfy the needs of the average householder. Instead, the main advantages are in the cost of maintenance. Quite a number of studies have been carried out in recent years into the full-life costs of equipping and supporting networks of computers. These consistently attribute around half the total cost to basic formal and informal administration — tasks such as helping each other out, backing up data, rebooting damaged software and so on. If one adds in system enhancements — configuration of new printers, loading of software upgrades, etc. — then annual figures as high as $10 000 are not unrealistic. Moreover, it also consistently emerges that PC LAN networks are more expensive to operate than those involving simple ('dumb') terminals attached to mainframe computers. The cost ratios measured range from twice to *15 times* as expensive to process a transaction on a PC network.

The concept of the network computer is intended to restore the balance of costs, by returning to the central administration the controls and tools necessary for ensuring that each PC on the platform has a set of software automatically available to it that is of the correct version and consistent with that of other users. Central control and distribution of software should, in principle, mean that company-wide roll-out of new applications can be achieved at minimal cost or disruption. This vision, however, remains to be tested and there are a number of commercial and organisational challenges that also need to be overcome.

LOCAL AREA NETWORKS

Unlike domestic terminals, the business machine is not normally connected directly to the wide area network. More usually it is part of a local area network (LAN) that serves the whole building or even the site. As described in Chapter 2, today's LANs have an architecture based on the limited reach and bit-rate available over copper, that is sufficient to meet the input and output speeds of desktop machines and their centralised servers. The most common standard employed is 'Ethernet', which allows individual terminals to transmit or receive bursts of data at rates of up to 10 Mbit/s. Machines share this 10 Mbit/s capacity: if more than one is transmitting at a time, then a 'collision' is detected and both have to back off for a random interval. Thus, on a busy LAN, the true throughput per terminal is considerably less than 10 Mbit/s.

Even 10 Mbit/s is not a particularly high speed by today's standards and represents a bottleneck that designers have been trying to overcome. Standards are being evolved for wired LANs that operate at much higher speeds. (Gigabit per second LANs are currently being talked about.) At the same time, there is a desire to cut costs in the fitting, removing and general maintenance of the connections to the network, by moving away from coaxial cable to simple unscreened pairs of wires. Instead of every terminal being connected to the same pair of coaxial cables, each is given a dedicated pair of wires that runs to a centralised switch – the original 'connectionless' concept of the LAN has been replaced by a switched (or 'connection' based) design. This makes the full 10 Mbit/s available to each terminal, irrespective of the simultaneous demands of the other terminals (provided the central switch can cope).

Note that the 10 Mbit/s for each terminal is shared between its sending and receiving capability – an attempt to do both gives rise to a collision. A simple modification, 'full duplex Ethernet', employs four wires per terminal, two for transmission and two for reception, thus effectively doubling the speed.

While cable-based LANs are continuing to push forward in terms of speed, two other alternatives are also emerging: radio and optical fibre.

RADIO LANs

Radio LANs have the obvious advantage that no hardwired connection is needed, leading to greater flexibility in office layout and the ability to work with mobile terminals. The market for them in 1994 amounted to around $50 million, with a projected increase to $250 million by the end of the century. The aim is to provide convenience rather than exceptional speed, with data-rates of the order of 2 Mbit/s, with a coverage within a building of around 10 000–20 000 square metres and the ability to roam between different radio access points. Recently, there has been some effort to standardise on a set of principles that will make interoperability between different vendors' systems a little easier. The standardisation is intended to allow a variety of different transmission techniques, notably 'Direct Sequence Spread Spectrum' and 'Frequency Hopping Spread Spectrum' (we shall explain these later) radio systems operating at 2.4 GHz, as well as infrared optical systems, to provide the same kind of service. As shown in Figure 5.13, the intention is that standard portable equipment will be equipped with plug-in radio units which will communicate with access points on the LAN. The access points create 'microcells', in a scaled-down version of a cellular network. If the mobile equipment moves out of range of access point A, then it re-registers with point B,

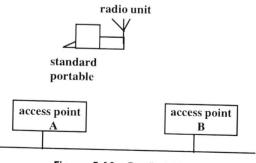

Figure 5.13 Radio LAN

and the communication continues. The standard does not lay down how the access points communicate to each other; this is left to the vendor.

Nor does the standard take a view regarding the two transmission techniques, Direct Sequence Spread Spectrum (DSSS) and Frequency Hopping Spread Spectrum (FHSS). Both of these techniques are intended to get round the problem of interfering signals that may well occur in the electrically noisy environments of modern offices and factories, while at the same time trying to keep the signal level of the radio LAN from further contributing to that noise. In the case of DSSS, the signal is 'spread out' across a wider bandwidth than is required to code up the data in a simple, conventional scheme; this makes it possible to recover the original signal, even if parts of the frequency spectrum are too noisy.

To understand the basic principles of DSSS, it helps to look at a rather simplified version of how the signal is coded. In Figure 5.14, we see a stream of digital data, which can be fitted into a frequency band A, say.

However, let us instead take the signal and transmit identical copies of it, within two other frequency bands B and C. We transmit the original and the copies in exact synchronism with each other. When we receive them, we convert them out of their individual bands and now have three identical signals, with exactly the same start and end times. We now add them together, and end up with a signal that is three times as big.

Let us now look at what happens to the background noise. There is

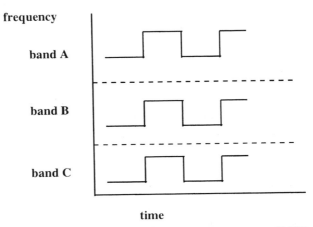

Figure 5.14 Direct sequence spread spectrum (DSSS)

a vanishingly small probability that the noise in each of the three bands will (a) be identical, and (b) begin and end exactly in synchronism with our wanted signal. So, when we add the outputs of the three bands together, the noise will not add up to three times its original value. Sometimes the noise in one band will be large and positive in value while that in another will be large and negative; sometimes they will both be nearly zero. For fully random signals, we can show that the noise will not grow by three times; instead it will grow by the square root of three, that is about 1.7 times. So, we have gained a 'noise immunity' of 3/1.7 by using the extra frequency bands. In general, the spread spectrum gain approaches the square root of the number of bands we use. Typically a spreading ratio of 10 is used, corresponding to a gain in immunity of just over 3 times (= square root of 10).

In a real DSSS system, although the noise immunity principle is exactly the same, the data is not sent as multiple frequency bands of the same signal. Instead, each data 'bit' is coded as a set of N bits, each of which is $1/N$ of the duration of the original data bit. This means that the original data bit is 'smeared out' (or 'spread') across a wider frequency range (N times greater) than it would be if sent as a single bit. The principle is identical to code division multiple access (see frame in Chapter 4).

Frequency hopping is rather easier to understand (Figure 5.15). The diagram shows two separate channels of data being transmitted simultaneously. This is done by each channel 'hopping' from frequency band to frequency band according to a pseudorandom sequence. (Sequences are chosen so that the two channels never hop to the same spot at the same time.) This means that the signal can hop around any loud narrowband signal for most of its time. Typically,

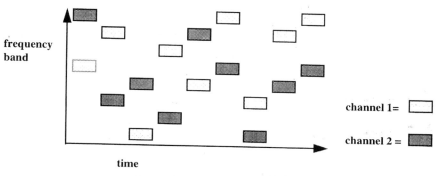

Figure 5.15 Frequency hopping

more than 75 frequencies are used and hops occur at a rate of 400 ms or faster. Since the systems operate at 1–2 Mbit/s, this is equivalent to around 1 Mbit of data per hop.

One of the reasons that standardisation has been attempted across the two radio methods is that proprietary products were available for both technologies. Consequently, there is vociferous argument between the commercial proponents of DSSS and of FHSS. FHSS is a slightly older technology and may be currently offering more concurrent channels. However, DSSS is seen by some as the longer-term solution.

There is also a problem in Europe with the choice of 2.4 GHz as a standard frequency, because this band is already overcrowded (although approval has been given in the UK). Perhaps as a longer-term solution to this problem is the emerging HyperLAN standard, which will operate within the microwave bands of 5.7 GHz and 18 GHz. Apart from being less crowded, these frequencies offer the possibility of 20 Mbit/s performance. This may become necessary, in order to compete with the high-speed Ethernet cable schemes that are becoming increasingly common.

OPTICAL LANs

Optical technology can be used to create two totally different LAN solutions: by means of optical fibre, one can create enormously fast, 'wired' networks; alternatively one can create the optical equivalent of the free-space, radio network, taking advantage of the greater bandwidth available through the use of optical frequencies.

Optical Fibre

Optical fibre is already quite widely used as the backbone for carrying traffic between switching hubs within buildings or across campuses and some organisations already deploy fibre to the individual workstation, where higher bandwidths or longer distances are required than can easily be handled by wire pairs or coaxial cable. At least one mature standard exists and is widely adopted. Fibre to the workstation is, however, relatively expensive, because of the low-volume production methods used in the manufacture of the transmitters, receivers and connectors. It is likely that fibre will become more attractive once the quest for much higher data rates has seriously begun. To give an idea of what might be achieved in future years,

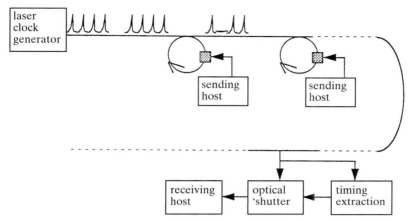

Figure 5.16 An ultrafast optical LAN ('SynchroLan')

we show, in Figure 5.16, one experimental design, the BT Labs 'SynchroLan'.

The laser source generates a very rapid train of clock pulses, say 40 billion pulses per second. These pulses travel through loops of fibre at each sending host where selected data pulses are modulated with data from the sending host. (In the diagram, the first 'sending' host has modulated the second pulse, reducing it to zero.) After modulation of the pulses by the corresponding sending hosts, the fibre LAN loops back through the receive part of each host. The receivers tap off some of the optical signal, extract a timing clock from it and, using the clock to drive the optical shutter, receive only the intended pulse.

The system is essentially very simple; the only optical pulse generation is by the source laser, which also generates the master clock. Thus, in principle, the design should be inexpensive to implement. Laboratory setups have already demonstrated the capability of supporting up to 16 nodes, at 2.5 Gbit/s each.

FREE-SPACE OPTICAL SYSTEMS

Current free-space optical systems rely on low-cost infra-red transmitters and are specified along with radio LANs to offer bit-rates of the order of 2 Mbit/s.

However, optical systems are capable of offering far greater bandwidth and it is possible that we shall see systems emerge which make use of the continuing reduction in the cost of higher performance components.

Optical propagation can be either multidirectional (diffuse) or line of sight. Diffuse systems can support full mobility, since a number of transmitters can be arranged to give total illumination within the room. The multipath reception results in a limited data rate × distance product which is unlikely to exceed 260 Mbit m/s, but for a typical size of 10 m × 10 m × 3 m this can still support 16 Mbit/s. For a higher rate, additional wavelengths of light could provide additional channels.

One important issue is the safety of the optical system. At 850 nm, for instance, the safe level is only 0.24 mW and this would mean that a large number of spatially separated devices would be required, unless we can ensure that the light from a single source can be diffused over a large area. One problem with this is that the laser sources required are essentially point sources and we need a simple form of diffuser that can spread this out in a reliable manner. One technique is to interpose a hologram between the laser and the rest of the room. Holograms can be etched into glass substrates, in the form of variable thickness patterns. Light passing from the laser into the hologram has its phase altered differently depending on the local thickness of the pattern and thus waves constructively and destructively interfere, converting the point source into a diffuse one. Although precise manufacturing of the substrate is required, it can be used to 'master' a large number of copies in plastic material, by simple embossing.

One alternative is to use an optical leaky feeder, which radiates along its length. This system is particularly suited to broadband, one-way distribution within buildings. The feeder would be installed along a ceiling or wall, perhaps in a transparent duct. One implementation is shown in Figure 5.17. This uses a D-section of optical fibre, produced by removing a portion of the circular cladding at the fibre preform stage, leaving the core close to the flat. When the fibre is drawn out, the shape is maintained and the core lies within a few microns from the surface. The fibre still maintains a low loss, guided

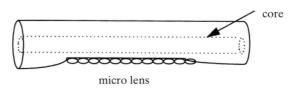

Figure 5.17 Optical leaky feeder

light property, unless a material of higher refractive index comes into contact with the flat surface, in which case light is guided into it, forming an optical tap. These optical taps can be created by microlenses to cover wide or narrow angles, according to requirements, and devices have been fabricated that can easily illuminate desk areas 4 m wide from a ceiling height of 2.5 m.

THE TRULY NETWORKED COMPUTER?

So far, we have considered ways of making the individual workstation communicate at higher and higher speeds with the external environment, consisting mainly of other computers. But if effectively infinite bandwidth becomes available, as is claimed by some, is it possible to consider the individual machine to be nothing other than part of a huge, worldwide 'supercomputer'? Looking at it another way, can we say that delivery of infinite bandwidth connectivity will destroy the concept of 'localness' in terms of computing architecture? The answer has to be 'only partially'. Why? Consider two computers connected by such an infinite bandwidth link, over a distance of x metres (Figure 5.18). At time t_1, the first machine issues two identical commands, one of which will be executed locally while the other is carried out on remote machine 2, which will return the result to machine 1 as soon as it has completed the operation, except for a delay introduced by their separation. In order for the two machines to be considered part of one local, multiprocessing computer, the time taken for the command to go from 1 to 2 plus the time taken for the response to return from 2 to 1 must be significantly less than the time taken to execute the instruction.

Today's computers can manage speeds of the order of 100–200 million operations per second. Taking a word length of 32 bits, this is equivalent to a data rate of a few Gbit/s, which is fast, but possible, at a price, over the optical wavelength division multiplex systems de-

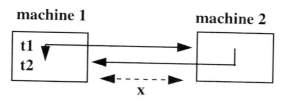

Figure 5.18 Networked computers

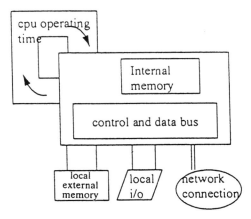

Figure 5.19 Hardware elements in processing operations

scribed in Chapter 3. But this is not the end of the story; even if the network can handle that capacity of traffic, it can still transmit it only at the speed of light, which in optical fibre is around 200 000 km/s. Thus, even a one-bit message will take as long to go there and back between machines 1 and 2 that are 1 km apart as the machines take to execute several hundred instructions. Since the world is about 40 000 km in circumference and computer operating speeds are still doubling according to Moore's Law, it is clear that infinite bit rate alone will not destroy the concept of locality in computing.

However, most computer tasks are not simply the execution of internal instructions. Figure 5.19 outlines some of the hardware elements commonly involved.

CPU operating cycles are only part of the story: invariably the program under execution requires a considerable degree of access to internal memory, usually significant access to local external memory such as hard disks, access to local input and output devices, and a network connection to services on a LAN or over a wide area.

Consider internal memory: the emphasis has been on the design of larger and larger memory chips, with much less success on improving access speed. Although processor speeds follow Moore's Law reasonably well, increasing between 50% and 100% each year, local memory speeds have increased only 10% per year over several years. Memory access is thus falling further and further behind progress in processor speed, with today's processors capable of executing several hundred instructions for each memory access.

The case is even more acute for external memory such as hard disk technology, which is even more concerned with capacity rather than speed. Disk accesses are usually restricted to continuous rates of a few megabits per second and have significant latency, of the order of milliseconds, in initiating data transfer.

What we also need to consider are the software protocols used for accessing these external resources. In Chapter 2, we discussed how existing Internet protocols are bad at handling very high speed data travelling significant distances, because they insist on acknowledgements from the receiving machine of blocks of data. The protocol issue is probably soluble in principle, although there are some problems in implementation, but the biggest practical problem may be in the need to support legacy systems originally designed for megabit, or less, bit-rates.

So, in trying to answer our question regarding the decline in the concept of 'local' versus 'remote' (or 'distributed') computing, we conclude that, for sheer processing power, local solutions will not lose their advantage (in fact, may increase it) for at least a couple of decades, but wherever significant access to memory is required, the local/remote distinction becomes vulnerable to the implementation of high-throughput networks. A few megabits per second delivered to the desktop are sufficient to match the performance of disk memory; internal RAM requires gigabit per second rates.

Increasingly, most applications require access to very large volumes of data, particularly where multimedia information is involved. Thus the major bottleneck is with external storage, usually hard disk. Once the 'information superhighway' begins to deliver a few megabits per second, then the local versus remote distinction can truly be said to have been seriously challenged.

ATM COMPUTERS

We have observed a tendency for LANs (and computer wide area networks) to move from 'connectionless', collision-based designs towards connection-based ATM services, where an end-to-end path with a defined quality of service is specified. It is interesting to speculate whether this might also influence the internal designs of workstations. Figures 5.7 and 5.19 show that they contain a control and data bus, which connects together the computing unit, the various memory and peripheral elements and the communications ports. It is possible that, in future, this bus could be structured on ATM lines.

This would not necessarily lead to any speed improvements – it should always be possible to produce buses which were optimised to single-machine and short-distance operation – but it would have one big advantage: it should be possible to allow a machine that required more memory or processing resources than it had locally, to set up a virtual channel to a nearby machine, without interrupting any other network activity.

Experimental implementations of these 'Desk Area Networks' have been created notably at Cambridge University Computer Laboratory. In their system, compressed data from a surveillance TV camera equipped with an ATM interface is transmitted across a network to a computer which can display the real-time, decompressed image and also carry out signal processing tasks such as motion detection. If motion is detected, then the relevant data can be sent to a hard disk. All of these processes – coding/decoding, motion detection and storage on disk – operate at variable rates. By using an ATM architecture, the experimental system can be optimised to handle network, computer bus and processing demands. It is possible to provide quality of service bounds on delay, variation in delay (jitter) and data loss.

6
Intelligent Management of Networked Information

THE 'INFORMATION CRISIS'

It is clear that the major growth in network traffic is in data rather than voice conversation, but such is the phenomenal growth that we have to ask ourselves 'what is it for?' and, perhaps more important, 'what are we going to do with it?'. This is the 'information crisis', the worry that our ability to create the information is greater than our ability to handle it. The problem with data is that, at some time, it will have to be interpreted or 'understood'. There are cases, for example the simple passing of sales data from a store to a head office computer that calculates the week's turnover, where the interpretation of the data is not too onerous, at least for a fast computer. But sales data also can be interpreted to identify what is selling and what is not. It can be examined for trends; it can even be linked to weather information to predict future sales and to supply chain services to schedule the deliveries. To do so, someone or some machine needs to carry out a sophisticated analysis.

Moreover, these figures are much easier to interpret than other data sets: satellite or medical images, identification of people by their faces or voices. The imminent requirement for data interpretation covers not jut text or numbers, but also sound and image, perhaps even taste and smell.

INFORMATION IN SUPPORT OF DECISIONS

In Figure 6.1, we repeat a vision for the management of information that has been developed by the US Department of Defense as part of their DARPA I*3 architecture. Information begins with raw data that occurs in a multiplicity of forms: text and speech, moving images, etc., which needs to be converted into basic units that have been 'understood' – in military terms, a tank has been recognised, words have

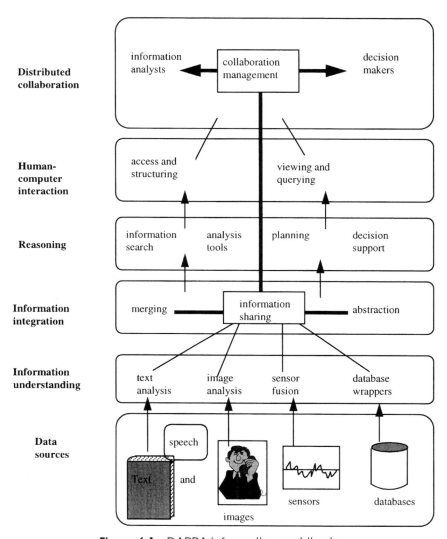

Figure 6.1 DARPA information architecture

been extracted from off-air intercepts of enemy telex, and so on. This information has to be categorised, filed, labelled with easy reference tags and merged with other sources, perhaps integrating the image with the text. The computers then have to carry out some simple reasoning on these sources of data: does the arrival of more images of tanks and an increase in the number of text references to 'fuel supplies' presage an unwelcome advance on a specific sector? If so, alerts must be raised by the system to the humans it supports and they must be allowed to query the information so that further data is presented in a way that is easy to understand, not cluttered by irrelevance, and structured in a number of ways, each to satisfy a task of a specific person. Finally, all of this must be made available across a network of analysts and decision-makers, again tailored to their specific purposes.

Whether the scenario is military, business, educational, healthcare or anything else, it is a tall order, but it is a reasonable target to aim for and in this chapter we shall look at how it might be achieved.

NETWORKED INFORMATION

As is common with many information architectures, the DARPA vision tends to imply that the user of the information – the military, in this case – has ownership and control of the basic data. Increasingly this is not often the case. There are problems enough with holding too much of your own information locally, but the problems are magnified when we look at multiple sources, created by different people on different sites across the world:

- Information is not created or stored in consistent formats.
- Seldom are there reliable pointers to where it is.
- It is not usually labelled with consistent classifications for rapid reference.
- It takes too long to download it.
- It takes too long to process it in the terminal.
- It takes too long to view it all at a speed where it can be understood.

Before we look at some of the ways that we can get round these problems, we need to consider the basic meaning of the term 'information'.

INFORMATION AND INTELLIGENCE

We have gone some way to explaining what we mean by information, in the previous paragraphs. Quite simply:

> **Data that is capable of interpretation, by human or by machine, is information.**

It is important to note that a string of 'bits' can be data to someone or something and information to someone or something else: it is doubtful that the tones generated by a telephone dial mean much to the average person but, to the network, they are meaningful items of information that can be used to route the call to its destination. On the other hand, to the network, the speech on the line may be simply meaningless data, however much information it might impart to the hearer.

In order to be able to interpret data, the person or machine involved has to deploy some amount of 'intelligence'. Unfortunately, intelligence is a very difficult concept to handle, and one which, when applied to machines, gives rise to considerable controversy. This is not the place to enter into the debate about whether computers can think or what is the true meaning of 'meaning' and other arguments. In the sections that follow, we shall refer to 'artificial intelligence' but we shall mean no more than the fact that a machine was able to deploy a stratagem which usually interpreted a chunk of data, for some purpose.

We do need, however, to look at least superficially at what is meant by the 'meaning' of data, in order to understand how it can be interpreted into information.

THE STRUCTURE OF INFORMATION

Philosophy is really only concerned with three deep-level problems: the ontological problem, 'What is there?', the epistemological problem, 'What does it mean?' and the moral problem, 'Is it good or bad?'. Leaving aside the third problem (which we do touch on, in Chapters 7 and 8!), we need to realise that artificial intelligent processing of information, in its own very crude and simplistic way, does address the first two profound questions, of being and meaning.

Philosophically, there are two extreme positions regarding the existence of 'things' and the meaning underlying them:

- We can believe that within the Universe there are things – dogs, cats, people, etc. – who are individually identifiable against a background of different things and who all share a set of attributes which taken together define a general property of dog-ness, cat-ness, humanity, etc., which is inherent in all dogs, cats or people and never shared across different categories.
- Alternatively, we can believe that there is no real form or order to the Universe. If we want to draw a circle round a set of points in the Universe and call the inside of the circle a dog, a cat, or even the class of all objects of the type 'dog', then we can do so, but we have merely defined a convention, not discovered a universal fact.

In most of our practical lives, we do not need to worry overmuch which, if either, of these principles is true, but when we come to design computers that are to have some sort of 'understanding', however little and however broadly we define it, then we have consciously to make decisions about how we define the 'ontology' (the entities) in our working space and the 'epistemology' (the meanings of them and their relationships).

Consider a simple but practical example, the design of a system to recognise faces. One way we could begin the design would be to define a set of rules for the things that make up a face (Figure 6.2). It will have one or two eyes, depending on whether or not it is in profile, but only one nose or mouth. It will not be too round or too long; we can set bounds on its 'ellipticity'.

By defining a set of things (or 'features') possessed by a face, we have then to design a system which can isolate these features from the data set.

Alternatively, we could assume nothing about faces, but simply collect a large number of examples and feed them (and perhaps also feed examples of 'non-faces') into some sort of statistical analyser which would eventually begin to recognise the difference between the data from faces and that from 'non-faces' (Figure 6.3).

This 'black-box' analyser will have developed a purely statistical

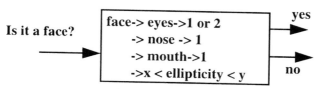

Figure 6.2 The rules of 'faceness'

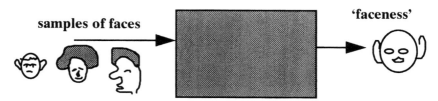

Figure 6.3 'Faceness' is just a statistic

view of what comprises a face. There may be no obvious connection between its idea of 'faceness' and the individual features that we identified in the first example.

Most practical recognisers use a mix of statistical and rule-based techniques.

THE VARIOUS FORMS OF INFORMATION

Quite separate from the intellectually challenging discussions regarding how information is structured, is the practical problem of how it is represented. Specifically, is it text, speech or image? Keyboard text is the easiest to deal with: the fundamental features (in this case, the letters, words, sentences) are already there and do not require extraction. Compare this with the problem of taking a speech sample that consists of a continuous, analogue signal that varies from speaker to speaker and from utterance to utterance. This is also true with handwritten text, which exhibits the same variability. At least speech and handwriting have one thing in common: we know what they are supposed to represent – words in a known vocabulary. This is simpler than the case of image recognition. What is the 'grammar' of an image, the equivalents of the 'parts of speech' that we find in text? Does such a thing exist or do we simply make up the rules to suit our current scenario?

PROCESSING STRATEGIES – POWER VERSUS INTELLIGENCE

The linguist Sapir once described language as a dynamo that could power an elevator but was usually connected to a doorbell. He argued that linguists should concern themselves with the complexities of elevator language, aiming to acquire detailed and deep understanding of the meanings and representations of our communications.

However, it turns out that the most useful results of automatic processing of information have usually been in ringing the doorbell, that is, in getting machines to interpret much simpler situations, without much recourse to deep structure. Indeed, it could be said that we need to use a great deal of machine power to produce even a weak 'ting' from the doorbell – we have to shake the whole building because we cannot provide sufficient intelligence to understand how to use the switch. Methodologically, the successes that have gradually emerged in information interpretation have come about through an increase in computer power, rather than in the sophistication and 'intelligence' of the algorithms.

The interpretation of information is intimately bound up with theories of language. In the 1960s and 70s, particularly in the area of speech processing, the great hope was a breakthrough in the understanding of the 'deep structure' of language, the discovery of the universal rules of grammar that were believed to underlie the way we all learn language as children, that could be copied into our intelligent machines to make them not just *recognise* speech, but even *understand* it. It is probably true to say that this has never happened. Theories of human grammar have undoubtedly developed significantly, but only a very limited amount of this theory has trickled down into machine intelligence. Machines recognise things by virtue of brute force searching of masses of data, using simple rules, rather than by understanding them through highly intelligent processing.

The artificial intelligence community know this and the search for real understanding continues. In the meantime, however, it is surprising what can be achieved by brute force.

INTELLIGENT PROCESSING OF TEXT

There are often just too many words. Often it is impossible to read through every detail of an article in the time available. For some years now, a limited amount of artificial intelligence has been available to those seeking out information, in the form of keyword searches. Individual words, or parts of words, linked together by AND, OR and NOT operations provide a reasonable filter on text files. Unfortunately, as any user will know, they are not particularly effective on large data volumes such as are available over the Internet and their effectiveness can only diminish, perhaps dramatically, as these volumes inevitably increase.

The trouble is that the intelligence of keyword systems is very limited: they only find a march of two things that are exactly alike and they take no real notice of context or the surroundings of the words.

CLASSIFICATION OF NETWORKED DOCUMENTS

Rather than look at the 'microstructure' of information sources – their individual words – it makes sense to look for some way of characterising larger chunks of information – pages or documents – so that they can be placed within some knowledge framework, in a manner similar to the way books are ordered in a library. One way would be to insist that the creators of the documents describe them in a standard way, using standard terms. Information which is not part of a document, but describes it, is known as 'meta data'. Meta data can be embedded in electronic documents, such as Web pages, without interfering with the look of the pages, but in a form that allows it to be recognised and processed by systems that access it remotely. (Web pages are composed using a desktop publishing language called HTML, which includes a number of formatting and other commands which are not displayed on the screen when the page is viewed in the normal manner.)

A simple example of meta data is the 'Dublin Core' (see frame). This is one of a number of proposals suggested by information scientists and librarians, for the marking up of the sort of information that one would expect to find in a library: books, journals, reports, etc. The idea is sound in principle and in many cases works well in practice, but it has its limitations.

The first problem is that it requires cooperation from the originators of the data. They must agree to mark up the document with meta tags that contain the necessary information, and they must remember to make any necessary changes. They must also be persuaded to use a standard set of categories (or a system that can easily be mapped into the standard). One problem immediately arises: the set of categories that is appropriate for one application, electronic library design, for example, may not be so useful for another, such as data from geophysical satellites.

Another problem arises when we try to populate some of the meta tag fields, for example in the case of the Dublin Core, with things such as 'subject'. The problem is not that there is no standard way to classify

META DATA – THE 'DUBLIN CORE'

Libraries are increasingly having to come to terms with the fact that more and more useful documents are appearing in electronic form, available over the Internet, and presented to readers on computer screens using standard Internet 'browsers', that make use of the fact that documents are written in a common syntax such as HTML (hypertext mark-up language). HTML contains hidden 'tags' that are not themselves visible on the screen, but are, instead, used to describe how the intentionally visible information should be presented – how big the print should be, where paragraphs, tables, etc., are to be used, and so on. Meta data tags can also be used to give information that neither is visible nor affects the visibility of the on-screen data: instead, they provide information that can be used for some other purpose, for example as part of an information search.

One such example is the 'Dublin Core', a set of 15 meta tags intended to describe a library document: Title, Author, Subject, Description, Publisher, Contributor (e.g. translator), Date, Type (e.g. a poem), Format, Identifier (e.g. an ISBN number), Source (e.g. from a book with ISBN number), Language, Relation (to other resources), Coverage (under development), Rights.

subjects. Rather, the opposite: there are dozens of classification schemes, each optimised for one set of applications but often either too detailed or not detailed enough for another.

Nevertheless, the discipline of defining a set of meta tags is a generally useful idea. What is being done is the identification of the significant entities of the application under consideration. We are effectively defining the contents of our universe, in terms of what is useful for us to recognise. Recently there has been a further development in the area of meta-tagging, with the emergence of a possible successor to HTML, known as XML. In Chapter 9 we discuss this further, in the context of a specific application scenario.

Although this is not always easy, it is certainly easier for us to do this than to develop computer systems that can themselves identify the significant entities. It is the human being, rather than the computer, that does the understanding. However, this is expensive and time-consuming and requires cooperation over significant periods of time. In the sections that follow, we shall look at some of the rather rudimentary, but often highly efficient, ways in which some of the intelligent understanding is done by the computer.

SUMMARISING TEXTS

Often the problem is not so much in finding a document, but in having time to scan through it to find out if it contains anything of value. What we would like is a simple means of producing an automatic summary, which captures the essence of the text, in much shorter form. What would be useful would be an automatic precis machine that could select just the key sentences for us. What do we mean by 'key'? Clearly there are a number of possible definitions, some requiring deep interpretation of meaning, others more simply examining the statistical relationships between bits of the text. Let us look at an example of the latter:

- Suppose we have a text consisting of a number of sentences.
- Take each sentence in turn and compare it with every other sentence in the text.
- Sentences are said to 'resemble' each other when they have, say, two words in common. (Words such as 'the', 'a', etc., are ignored.)
- The sentences with the highest number of resemblances are assumed to be the 'key' sentences.

For example, consider the text, nine sentences in length, shown in Figure 6.4. We can rank the sentences in terms of the number of 'resemblances', i.e. links that they have with each other:

- Sentence 6 resembles 1, 2, 3, 4, 7 and 9 (six links)
- Sentence 4 resembles 1, 6, 7 and 8 (four links)
- Sentence 8 resembles 3, 4 and 9 (three links)
- Sentences 1, 3, 7 and 9 have two links
- Sentence 2 has only one link
- Sentence 5 has no resemblances

This process creates a ranking table of the sentences in terms of their typicality to the text. For any desired degree of summarisation, we just choose, from the top, those with the greatest number of resemblances, until we reach the desired amount of text.

This summarisation technique could be used either to display a summary of each of many passages we wished to search, on the limited space of a screen, or to act as a prefilter to a keyword search.

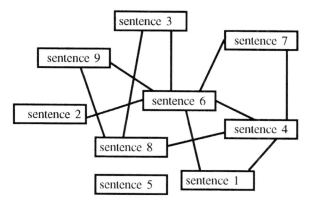

Figure 6.4 'Links' between sentences

AUTOMATIC CLASSIFICATION OF DOCUMENTS

What makes a book about dogs different from a book about cats is that the former will use the word 'dog' a considerable number of times and the word 'cat' infrequently, if at all. It may also say that 'Dogs chase cats', but it is unlikely to say that 'Dogs like to catch goldfish'. A cat book, on the other hand, may well refer both to dogs and to goldfish. The linguistic significance of the dog and cat books lies partly in the frequency with which they refer to 'home' terms, such as 'dog' in a dog book or 'cat' in a cat book, and partly to the frequency with which they refer to 'away' terms, such as 'cat' in a dog book, 'dog' in a cat book and 'goldfish' in either book. Texts are significant, not only because they share certain words, but also because there are words that they do not have in common. We can use this as a way of classifying documents.

Suppose, for example, we wanted to summarise the whole of this book. It would probably not be very meaningful to apply the 'two words in common' rule that we used in the previous section, to every sentence and then rank them in order. Why not? Simply because, like most long texts, the subject matter changes from chapter to chapter. What would be better would be to summarise each chapter individually.

Now, this book follows the convention of most books in that chapter headings are clearly identified. But suppose we (or our computers) were ignorant of this convention, or suppose we wanted a computer to segment a long chunk of text up into units that behaved like conventional 'chapters'. How could it do so? Essentially, what makes chapters into chapters? Simply the fact that they are on different subjects. Instead of concentrating on words that are common across a number of sentences, we can look for words which occur more frequently in restricted parts of the document than they do elsewhere. For example, in Chapter 3, 'Advanced optical fibre networks', there are approximately 80 references to 'fibre', but there are only three in Chapter 4, 'Radical options for radio' (where you might indeed expect some such reference to fibre as a competing or interworking technology), and none in several other chapters.

In general, we can classify texts as 'distinctly different' from the average if the Term Frequency Inverse Document Frequency (TF-IDF), defined as:

$$\text{TF-IDF} = \frac{\text{usual frequency of occurrence of a set of words}}{\text{frequency of these words in the document}}$$

is small.

Obviously, we can nest these degrees of internal consistency: although the chapters in this book vary in their detailed contents, they all use the word 'telecommunications' more frequently, and the word 'goldfish' much less frequently, than would a whole library of books on pets or zoology.

MACHINE TRANSLATION OF TEXT AND THE 'FRAME PROBLEM'

One of the early expectations following the development of the computer was its use as a means of artificial translation. All it needed to do would be to look words up in a bilingual dictionary and replace one with the other. That this did not take place is obvious to all; why it did not happen is less easy to see. The problem is not just one of 'language'; it strikes deep at the heart of artificial intelligence. Thus it is worth looking at, not just because translation is an important requirement in an increasingly global information environment, but also as a way of understanding what we can and cannot expect from computers, given our current state of knowledge.

Consider the following text:

> *The problem with transmissions of multi-media is that telecoms that the companies can provide guaranteed the bandwidth at the low latency (low instability of delay) and (variation of delay), but at the excessive prices. In addition, the data-processing networks based on the Internet protocols*

INTELLIGENT MANAGEMENT OF NETWORKED INFORMATION

and the practical installations of the under-nets and the limited servers of field of flow, cannot promise any quality of service beyond the delivery of package, the major part of time, thereafter. Distributed architectures of computer often discuss the functionality but not the execution. They are particularly bad with the ordering of flow.

It is just about readable and seems to convey a fugitive sense of meaning. But what do we make of phrases such as 'of the under-nets', 'low instability of delay' and 'with the ordering of flow'?

The text is, in fact, a translation from French of a passage that was in turn the result of translating an original English text, i.e. English → French → English, using a commercially available machine translation service. The original is as follows:

The problem with multimedia communications is that the telecoms companies can provide guaranteed bandwidth at low latency (delay) and low jitter (delay variation), but at excessive prices. On the other hand, computer networks based on the Internet protocols and practical installations of subnets and limited throughput domain servers, are unable to promise any quality of service beyond packet delivery, most of the time, eventually. Distributed computer architectures often discuss functionality but not performance. They are particularly bad with flow control.

We have to congratulate the machine on achieving a reasonably grammatical solution and good translation of the greater part of the common words. (In a sentence generated by a human writer that would not itself win prices for clarity.) It has, however, had serious problems with words which are not themselves exotic, but are used in a specialised way to convey technical concepts. The mental confusion we experience when reading the translation comes about because a few, but critical, parts of the text are not effectively dealt with. What is happening is that the software is good at 'structural details' but much poorer at 'semantic' elements. The problem is that 'meaning' resides largely in significant differences between texts, as we saw when we looked at Term Frequency Inverse Document Frequency (TF-IDF) in the section above. These differences represent the special, localised viewpoints that people adopt when they work on specific tasks. We are predisposed to recognise words, even badly spelt or spoken ones, that relate to the 'domain' in which we are currently thinking. We create, from fuzzy outlines, images that match what we expect to see. But we are also able to jump from one viewpoint to another, provided circumstances have changed at a higher level. We have not been very

FROM CHEESE TO GLUE

The following Italian text was translated to English and back, several times:

> *Fabrizio MIl formaggio di Sardegna. In Sardegna, nonostante l'avvento delle nuove tecnologie, la lavorazione del formaggio procede sempre secondo le regole tramandate. Il latte viene scaldato e si aggiunge il caglio in attesa che si condensi. Si sgocciola in una forma e viene poi messo in salamoia per la stagionatura. Quello che rimane viene utilizzato per la produzione della ricotta. Al formaggio della Sardegna e' stato riconosciuto il marchio DOC, nelle diverse espressioni a seconda delle zone di produzione. Alcune di queste sono: il Fiore sardo e il Pecorino sardo. Il Fiore Sardo e' costituito da Pasta cruda che viene affumicata tramite procedimenti tradizionali dopo un certo periodo di salamoia. E' particolarmente buono nel primo periodo da tavola, ed in un secondo tempo puo' essere grattugiato ed utilizzato, per esempio nel condimento dei primi piatti. Il Pecorino sardo e' costituito da una pasta piu densa e piu' fortemente aggregata. Ben si presta ad essere consumato a tavola nel primissi*

resulting in the following English version, which certainly does not sound appetising. The emergence of the 'glue' appears after three passes through the system, but clear errors occur after the first pass:

> *The cheese of Fabrizio MIl of the Sardinia. In Sardinia, in spite of the event of new technologies, the operation of the cheese in those second ones place continues has always passed on the rules. The latte ones come scaldato and the rennet helped to is associated to how much is condensed. Sgocciola in a figure and it then has come has put in pickling brine of marinatura of the marinatura of the marinatura for the stagionatura. That those remains comes has used for the production moderated to the quiet. To the cheese of the Sardinia and 'the document of recognized of it the marks, in the several expressions to those second ones of the production zones. Some of these are: the Sardinian flower and the Sardinian Pecorino. The beginning Sardinian and' constituted of the flower from the crude glue that comes is filled up in on within on smoke of the traditional procedures after a sure period of the marinatura of the pickling brine of marinatura of marinatura. Especially buoa and 'in the first period from the table and to the inner part of*

successful in transferring these skills to computers. This is one variation of the 'frame problem' of artificial intelligence: we can specify rigidly the domain in which they should operate, in which case they can perform adequately but cannot change to a new viewpoint, or we can allow them to roam freely across an information space, in which case they end up exploring an impossibility large set of options and overlapping possibilities.

In the case of our machine translation example, we could improve the performance by specifying that the domain of knowledge was that of 'communications and computing', thus allowing the computer to select a dictionary which translated 'subnet', 'flow control', etc., into their other-language equivalents, but this would have meant that a human might have been necessary to make that decision in the first place. This is particularly true when the domain of knowledge changes throughout a text. This is even worse, as we shall see later, when we deal with variably produced material, such as speech.

That said, automatic translation does provide a limited amount of assistance to users who do not understand the original. Machine strategies are being continually improved and, within tightly specified domains, it is likely that we shall see a gradual introduction of translation support tools over the next few years. In the meantime, a further example, hopefully both amusing and instructive, of the current limits of machine translation is given in the frame 'From cheese to glue'.

UNDERSTANDING SPEECH

Speech, of course, holds a rather special place in telecommunications and it is still by far the largest single message type carried by the network. It also has special importance in the control and interpretation of other sources of data: if we could simply talk to machines a major barrier to many network applications would be overcome. Similarly, it is one of the most natural and emotionally satisfying ways of receiving information. But despite this need and many years of research, we are not yet surrounded by machines that listen to us and can understand what we say, or speak back to us in a natural and sensible way. In this section we shall look at the problems of automatic speech recognition and synthesis, and some of the progress that has been made. A detailed description of human production of speech is beyond the scope of this book, but a brief outline is given in the frame 'Elements of speech production'.

> ## ELEMENTS OF SPEECH PRODUCTION
>
> Speech sounds are generated as a result of one or both of two sources of sound: a 'voiced' source generated when air expelled from the lungs makes the vocal cords vibrate rather like the reeds of a musical instrument, at a regular frequency or 'pitch' (which can vary over 2–3 octaves), or an 'unvoiced' source generated by friction as the air passes over obstructions such as the teeth or the lips. What is heard from our mouths is these sources, modified by the natural resonances of the vocal tract, from larynx to lips. For example, different vowel sounds are formed by using the position of the tongue to create a constriction between it and the roof of the mouth, at different places:
>
>
>
> We can consider the cavity behind the high point of the tongue to have a resonant frequency of F_1 hertz and the cavity between the tongue and the lips resonates at F_2. As we move the high point of the tongue forwards, the resonant frequencies change, approximately 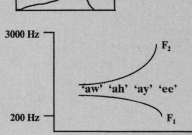 as shown in the diagram, passing through a range of vowel sounds. In detail, there are more than just two resonances (and, in the case of nasal sounds, some antiresonances). The precise values of the resonances vary from speaker to speaker, even from one utterance to another for the same speaker, and depending on the sounds that surround the vowel.

WHY IS SPEECH SO DIFFICULT TO RECOGNISE?

In the past there were power limitations in our processing capability that made it difficult to handle the task at sufficient speed, but this alibi is probably no longer viable. The problems are primarily a result of our limited theoretical apparatus for analysing the basic sound units in terms of what they are intended to represent. One major difficulty is in natural variability: although phoneticians tend to analyse speech into a string of 'phonemes', approximately the spoken equivalent of the written alphabet (there are about 50 phonemes used to describe spoken English), there is no simple, one-to-one correspondence between the phoneme and the continuously varying and variable speech signal. For instance, individual speakers vary from utterance to utterance, even when saying the 'same' thing. It follows that different speakers can generate even more dissimilar sounds when they use the same words. There are no simple isolated sounds in continuously spoken speech; there are not even, necessarily, boundaries between words. The contexts in which any nominal sound is set, particularly the sounds either side of it and the speed of the speech, also grossly affect the sound produced, as the articulators (tongue, mouth, etc.) adjust dynamically, 'undershooting' the target sound, for example.

Failure to produce the sounds in a consistent way is not the only problem. It is not simply that one intended message has many actualisations; sometimes more than one message is communicated in identical sounds: there is nothing to distinguish the usual pronunciation of 'To recognise speech' from the phrase 'To wreck a nice beach', except by reference to the semantic context in which they are set. It is only by reference to such contexts that we can determine what is probably meant in cases like these.

PRINCIPLES OF SPEECH RECOGNISERS

Whatever their scope of operation (isolated word/continuous speech) or range of users (single/multiple speaker), speech recognisers can be considered to consist of the major building blocks illustrated in Figure 6.5.

It starts with an electrical copy of an acoustic signal, the speech, which may, for example, come from a high-quality microphone in a

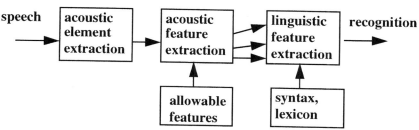

Figure 6.5 Speech recogniser components

quiet environment, or over a rather poorer-quality telephone line from a speaker in a moving car. Nevertheless, the principle is the same: the raw speech must be converted into some set of elements that are more amenable to the recogniser. A common example of this 'acoustic element extraction' that has been used since electrical processing of speech began seriously during World War II is frequency-band analysis: the speech signal is broken up into a number of contingent frequency bands and the energy in each band is averaged over a period of a few tens of milliseconds. Thus the speech is represented by an array of N by M numbers, where N, the number of frequency bands, is typically of the order of 20 for telephone-quality speech and M is the duration of the complete utterance (in seconds) divided by the averaging time of the energy (say 30 milliseconds). An extended passage of speech will then be represented as shown in Figure 6.6.

There are other sets of elements we can use, apart from simple frequency analysis, but the principle is just the same. We now need to convert these elements into 'features'. That is, we need to take them individually, or in groups, and reduce their infinite variety into a finite, manageable set of entities that are 'features' of speech.

An early technique was to use the whole $N \times M$ array to represent a whole word, or even a phrase. A standard 'template' is created for each word in the vocabulary and stored in the system. When the system is used to recognise a word, its array is compared with every template and the best match then determines the word.

How do we measure the 'match'? A number of methods are commonly used. For instance, in Figure 6.7, we show how to calculate the 'mean squared Euclidean distance' between a standard template and

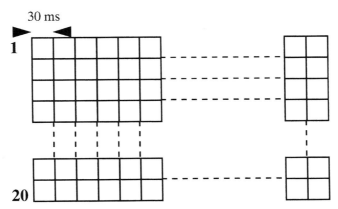

Figure 6.6 Simple array of acoustic elements for speech processing

INTELLIGENT MANAGEMENT OF NETWORKED INFORMATION

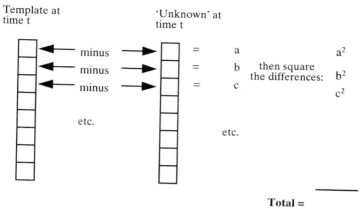

Figure 6.7 Mean-squared Euclidean distance

an unknown one, at a single instant in time. In the example, we have used only eight filter bands. We subtract the energy in each filter in the unknown from the corresponding filter in the template, square the result and sum all the squares. This gives us a measure of how different the two patterns are, at that instant in time.

We can now repeat the process for each pair of time samples from the template and the unknown and add together all the distances thus calculated, to obtain a measure of how dissimilar are the template and the unknown across their entire durations. We carry out this measurement between the unknown and all the templates. The lowest value corresponds to the most likely match between template and unknown.

There are distance measures other than Euclidean mean-square that can be used and distances can also be weighted to give more significance to elements that have been experimentally shown to be particularly good differentiators (in our case, we could weight certain frequency bands more than others), but there is a problem with this simple approach. Comparing each utterance with the template (Figure 6.8) assumes that the speed of uttering the sound in both cases is the same, so that cells at time t in the template should be compared with those at time t in the unknown utterance. This will almost never be the case, simply because the rate of speaking will vary from one time to another. Nor can we simply stretch or shrink all of the pattern by the same amount: we said earlier that the dynamics of the articulators result in different patterns for different speeds. In particular, some sounds, such as vowels and other 'continuant' sounds, change their duration by a higher percentage than do explosive sounds such as 'k' or 'd'.

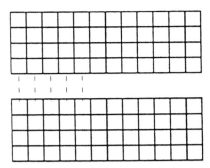

Figure 6.8 The time-registration template matching problem

The initial theoretical approach adopted to get round this problem was 'dynamic time warping' (see frame). This attempted to find the shortest end-to-end distance between the template and the unknown by lining up the closest samples, whether or not they occurred strictly at the same nominal time, successively finding the shortest route from earlier parts of the utterance to later parts. Not only did this allow for differences in speed, it was also an algorithm that could be calculated 'on the fly' rather than by having to wait until the complete utterance was sampled.

Important though this approach is in its own right, as a way of allowing variable samples of speech to be compared with a set of standard templates, perhaps its greatest significance was that it led to the investigation of the possibility of using flexible models based on the existence of 'hidden' features of speech, and the technique of 'hidden Markov modelling' (see below).

HIDDEN FEATURES OF SPEECH

In our examination of dynamic time warping, we saw that sometimes more than one time-sample of the 'unknown' utterance could be mapped into a single time-sample of the template (Figure 6.9). This is

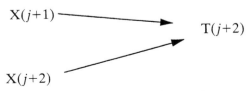

Figure 6.9 Hidden 'states' of speech

DYNAMIC TIME WARPING

Suppose we want to compare an unknown utterance 'X' with a template pattern 'T' both X and T are represented by a set of values (such as the short-term energies in a set of N bandpass filters), repeatedly

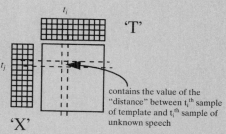

contains the value of the "distance" between t_i^{th} sample of template and t_j^{th} sample of unknown speech

measured throughout the utterance. Our problem is that we do not know which set of N samples of X should be compared with which set of N samples of T, because the differences in speed of speaking will mean we cannot simply line them up against each other.

We want to develop a method of stretching and squeezing the timescales ('time warping') of the two patterns, to get the best fit.

We start by calculating the distances (as described in the main text) between every sample in T and every sample in X. This results in the grid shown here which gives a complete map of the distance between every time sample of X and every time sample of T. Note that the grid need not be square: T and X do not necessarily have the same duration.

To help comprehension, let us first of all imagine that we compare template T with itself. Then the first set of filter energies will obviously be no distance at all away from themselves, the second likewise, and so on, all the way to the end. This will result in the grid having zero everywhere on the diagonal, as shown.

In general, the off-diagonal terms will not be zero and so the minimum total distance from start to finish (zero, in fact) is got by travelling all the way down the main diagonal. This is equivalent to saying that cell i should be considered to match up best with cell i (obviously), rather than with cell j, where j does not equal i. However, when the two patterns, X and T, are not identical, this may not be the case.

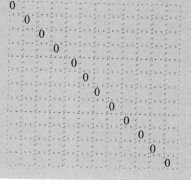

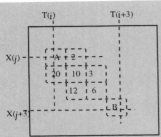

Consider the segment of the grid of distances shown in the diagram. Suppose we have successfully lined up template point T(i) with unknown utterance point X(j). For example, they could be the first points in each where the speech began. Let this be the point A. In going from A to B, the point T(i + 3) and X(j + 3), we could travel the diagonal, incurring a distance increase of 10 + 6 units. But an alternative, and shorter, route is to move off the diagonal, picking up distances 2, 3 and 6. This is equivalent to saying that samples T(i) and T(i + 1) both correspond to sample X(j) and that samples X(j + 1) and X(j + 2) both correspond to T(j + 2).

the germ of an idea that we should move away from looking simply at basic acoustic elements, such as the outputs of filter bands, towards the idea that these outputs are actually observations of a set of distinctive 'features' which in some way represent states of steady movement or rest of the articulators (tongue, lips, etc.) of the utterance. The precise observed behaviour varies – the filter outputs never have exactly the same value or duration each time the word or phrase is repeated – but there is an underlying structure to the speech. That is, of all possible sounds and combinations of sounds that could be generated by the human voice, there are a restricted set which together define a 'model' of speech production. If we can understand this model, we can know what to look out for, despite the random variation that also occur, between speakers and from utterance to utterance.

HIDDEN MARKOV MODELS

The idea, then, is to come up with a 'model', essentially a computer algorithm, which can represent speech as a (finite) set of features which underlie the production of each word, at the same time as allowing for them to have a degree of statistical fluctuation. One particularly successful example is the 'hidden Markov model'. In order to see how this can be used for recognition, it is better to look

INTELLIGENT MANAGEMENT OF NETWORKED INFORMATION 199

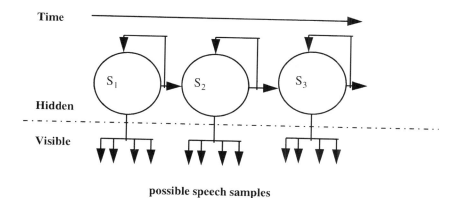

Figure 6.10 Hidden Markov model of speech production

first at how the model represents speech production. Let us imagine designing a machine that is to speak a particular word, not like the playback from a tape recorder, always uttering exactly the same sound, but rather like the way humans do it, with variation from time to time and from speaker to speaker. Consider the model in Figure 6.10. This example shows a three-state model of a word. Each of the states represents a speech 'feature' – we do not need to ask exactly what it is, because it is defined only by its statistical behaviour. (However, to make it clearer, imagine the state to be related to a restricted range of positions that the vocal tract can take up and still make approximately the same sound.)

Let's make up an example: suppose we have a vocabulary of two words, 'wait' and 'hate'. Then we could imagine both words to have three states each, corresponding to sound patterns 'wih', 'ay', 'tih' and 'hih', 'ay', 'tih', respectively. Let's choose a simple set of filter outputs as before. The filter outputs for the state called 'wih' will be, in practice, slightly different every time the word is pronounced – we've shown four possibilities; sometimes, indeed, it will be difficult to know whether we have a sample corresponding to 'wih' rather than to 'hih'. Also, the number of filter outputs corresponding to 'wih' will also vary depending on the speed of speaking. The states are thus represented by the statistical behaviour of two parameters. One is the probability (a number between 0 and 1) of the state being replaced, at the next instant in time, by the next state. This is the 'transition probability'. Suppose the first state in our word has a transition probability value of 0.3. So, if we use the machine to generate 10 examples of the word and we could look 'inside' the model, we would find that, three out of 10 times, the word was represented by a model whose first state lasted only one time step. If we looked at the other seven utterances of the word, we would find that about two of them (= 0.3 × 7) had changed state at the next time step, and so on. The transition probability provides a natural way of scaling for speed variations.

The second statistical property of each state relates to the sound it is associated with. We select some way of representing these sounds; let us say we use samples of the outputs of a filter bank, as before. In principle, we could have an infinite variety of samples, but, by quantising them into a finite set, we can make the explanation simpler. (This is often done in practice.) In Figure 6.10 each of the three states has been allowed to generate any one of four values. These values are generated at random, but they are not equally probably. Suppose for state S_1 we call its four samples 'a', 'b', 'c' and 'd' and assign probabilities to them of 0.5, 0.3, 0.15 and 0.05, respectively. We are now in a position to begin to operate our 'word machine':

1. Take a pack of 100 plain cards and write an 'a' on 50 of them, 'b' on 30, 'c' on 15 and 'd' on five. Call this pack X.
2. Take a pack of 10 plain cards and, on seven of them, write the word 'stay;'. On the other three, write 'move'. Call this pack Y.
3. Shuffle pack X and look at the bottom card. Suppose it is 'b'.
4. Shuffle pack Y and look at the bottom card. Suppose it is 'stay'.
5. Take the X pack, cutting again and getting a 'c'.
6. Take the Y pack, cutting and getting a 'move'.

So, for this utterance of our word, we have completed the first state and generated the first two speech samples – 'b' and 'c' – that we would use to drive our speech synthesiser. Pack Y, our transition probabilities, has now told us to move on to State 2, and we need to replace our X and Y packs by ones that reflect the statistics of State 2. Thus, we can see that the progress of the utterance of the word by our model can be represented by the passing from one state to the next, emitting speech sample $S(t)$ at time t.

Figure 6.11 shows the same thing for a more complex model. In this example of a five-state model, we see how the model evolves with time for a particular utterance of the word. In another utterance, the durations of the states might be different and the speech samples $S(t)$ will also be different.

That is all very well, but how do we know what values to give to the speech samples, their probabilities and the transition probabilities of

INTELLIGENT MANAGEMENT OF NETWORKED INFORMATION

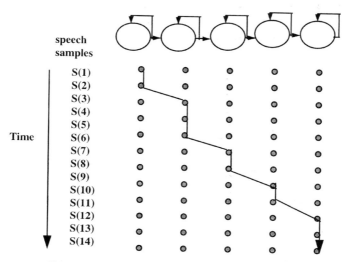

Figure 6.11 Five-stage hidden Markov model

the states? And how do we use the model for recognition, rather than production, of speech?

The answer is rather complicated and a full explanation requires a significant amount of mathematical detail which is inappropriate to this text. It is also true to say that the solutions, though mathematical, are not rigorous and there is a considerable element of *ad hoc* experimentation and craft skill. However, we can gain some idea of how it is done, without becoming too involved with the calculations.

First, it is worth reflecting on what we know, and do not know, about the 'true' underlying model of speech production for our words in question. Look again at Figure 6.10. Even if the model were indeed a true representation of how the speech was produced, we would have no way of knowing that, because the model lies hidden; all we have as evidence are the speech samples themselves. We do not need to strive to create a 'true' model – indeed it would be pointless, because we would never know if we had done so – all we need to do is to build one on the basis of the observable data that we have, the speech samples.

So, suppose we have a set of words we wish to recognise and a number of examples of each, spoken by a representative collection of speakers. Then, for each word, we decide on the number of states that we are going to allow them to have. We can use experience or simply a fixed number; too many will result simply in more calculations, too few in too little accuracy. We now need to calculate the statistics of each state.

We said that the individual states tend to be associated with individual speech features, deliberately without trying to be too precise about what a feature is. Typically, experimenters have used steady sounds such as vowels or common groups of sounds at sub-word level, such as 'atch' in 'catch'. Multiple examples of them are recorded from a representative population and their statistics are calculated.

In order to use the model as a speech recogniser, we now reverse the process that we used when we wanted to generate speech: in that case, we used the probability of a sound given a state; in the recogniser case, we calculate the probability of each state being responsible for the observed sound, and select the state with the highest value. (We look at two sets of probabilities – how likely is the sound produced to have come from any state, and how likely is the model to be in that state, at this time. The maths is quite complex, but the reasoning relatively primitive.)

We do this across all the states for all the words in our recogniser's vocabulary and choose the most likely.

The hidden Markov model has the advantage that it can also be applied to the linguistic, as well as the acoustic, aspects of speech (see Figure 6.5). Suppose we wish to recognise sentences rather than isolated words: we can use grammatical information in a model to account for the transition probabilities of verbs following nouns, for instance. In a specialist recognition domain, e.g. for technology, we could even have lexicon probabilities such as 'networks' following on from 'computer'. Combining these with sub-word recognition gives us a powerful statistical model across a number of different levels of analysis.

HOW GOOD ARE SPEECH RECOGNISERS?

Of course, in the end, the requirement for speech recognition is performance, rather than a theoretical model, and we need to ask 'How good are speech recognisers in practice?'.

Unfortunately, this is a rather difficult question to answer simply, because we have to consider their performance over an infinite range of possible sounds, an effectively infinite population of speakers and environments with greater or less background noise and, perhaps, distortion brought about by telephone lines or telephones. It is easier

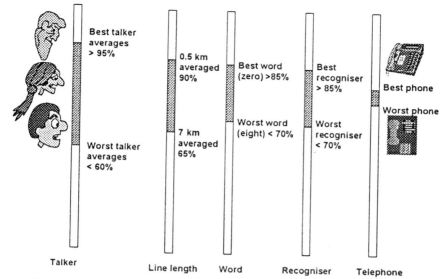

Figure 6.12 Performance of speech recognisers (source: BT Labs)

to carry out comparative testing, using a closed sample of words and speakers. The results of one study, carried out at BT Laboratories, are summarised in Figure 6.12.

The demands placed on the recognisers on test were modest: simply the ability to recognise isolated numbers from zero to nine. Even then, it can be seen that the results were by no means perfect. The single factor which had the greatest impact upon the performance of a recogniser is the speaker's voice. This was the case for all recognisers, although some were less variable than others. The telephone line was the second biggest factor with performance deteriorating as lines got longer. Words and recognisers came next with the telephone instrument being less important. This experiment also used two (relatively quiet) levels of background noise and found that these were almost insignificant. This does not imply that background noise is unimportant – it merely tells us that in normal quiet environments a *small* variation in background noise (10 dB) is less important than the other factors.

Dwelling a little further on the issue of the performance of systems with regard to individual speakers, we can examine some results that were achieved in the 1994 competition held by ARPA, in the USA. A very large vocabulary – 65 000 words – was tested on a number of competing systems. The performance of the winner is shown in Figure 6.13.

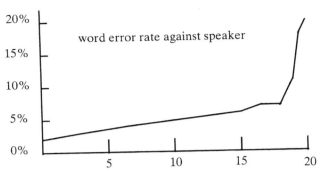

Figure 6.13 Speech recognisers – the ARPA test (Acoustics Bulletin, 1995)

The graph gives the average error rate for each speaker. It can be seen that, for 16 out of the total of 20 speakers, the error rate was quite low, of the order of 7%. However, for the remaining four, that is, 20%, the error rate was considerably higher. This inability to recognise data from a significant 'tail' of the test sample is a problem common to most speech (and speaker) recognition systems and may be a major problem to overcome if they are to become accepted.

DOMAIN RESTRICTION AND DIALOGUE DESIGN

The above results are encouraging rather than spectacular. Clearly, automated speech recognition is far from reliable, given an unrestricted domain of speakers and words. We know we can do better in some instances, not because of theory but because there are systems on the market now that can take audio dictation and produce reasonably accurate output for vocabularies of tens of thousands of words. These systems are not superior developments in terms of having a deep-level understanding of speech; they achieve their performance by restricting their domain of operation, usually by being speaker-specific and trained over a fairly lengthy period on the user's voice. Secondly, they make extensive use of the statistics of groups of words: they have been given large databases of the probabilities of the occurrence of individual words, probabilities of pairs of words occurring together, and so on. Some of them also make use of statistics tailored to the domain of discourse: if we are talking about financial matters, 'stocks and shares' is a more likely phrase than 'socks on stairs'. This is simply a further manifestation of the 'frame problem' discussed earlier in this chapter.

INTERACTIVE VOICE RESPONSE: A TELEPHONE ORDERING SYSTEM

A further twist can be introduced in the case of systems which carry out a dialogue with the user, for instance a telephone ordering service. The taking of orders or payments over the telephone now makes up over 50% of remote shopping transactions, corresponding to about 2% of total sales, exclusive of automobile sales, in the UK or the USA. However, providing 7 days a week, 24-hour answering capability using human call-centre operators is very expensive. One alternative is to use an automatic interactive speech system, which holds a list of many thousands of customers and their account numbers. Figure 6.14 describes such a system.

Although the system is not perfect at recognising either a name or a number, it can estimate the most likely values (and an estimate of how likely they are) of both of those and combine their probabilities to make a good guess at the correct answer. The system contains a branching structure of questions and responses between it and the user. If this dialogue is skilfully designed, then the number of legitimate responses the latter can make can be made very small, for example 'yes', 'no' or the specification of an account number or a

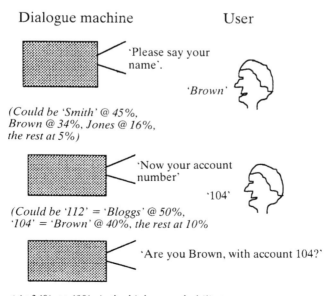

Figure 6.14 Speech recognition for automated order taking

listed name. This makes the recogniser's task much simpler at each decision point. In the event that the guess is incorrect, the user's reply ('no') is usually recognisable and the user can be prompted to try again. In the event of complete failure – which will happen sometimes, with some speakers – the system can simply pass these occasional failures to one of a very small number of human 'exception handling' operators.

SPEECH SYNTHESIS

In the interactive system just described, not only did the computer recognise the speech but it also spoke back. In the same way that speech input is one of the most natural ways to provide input or control to a machine, speech output from the machine is also the most natural and often the most convenient. In some ways the task is easier than speech input – for instance, the input signal, a string of text, does not possess the inherent variations of the speech signal – but there still exists a gap between the ideal solution and what we can achieve today.

The simplest way to generate speech sounds is to record them from a live speaker as complete phrases and play them back when required. This also creates the most natural-sounding message. By using relatively simple speech coding methods, it is possible to reduce the amount of memory required for storage to something of the order of 1 Kbyte for each second of speech, or less. Thus storage is not usually the limiting problem. The real difficulty is in the recording: you need to get the narrator to speak each and every phrase that you will require; how can you hope to estimate in advance all the changes that will be required through the lifetime of the system? It is not always possible to call people back for subsequent recordings: some people's voices change dramatically with time, some die, some simply demand too much money!

An alternative is to go for a completely artificial voice, synthesised from an abstract model of speech generation. Speech is synthesised by altering the various resonances in the 'vocal tract filter' (see the frame earlier in this chapter on 'Elements of speech production') and selecting a buzzing or hissing source to put through them. The repetition rate of the buzz is continuously altered according to the pitch profile required to make the utterance sound more natural, and the volume control is also changed to give the correct emphasis (Figure 6.15).

INTELLIGENT MANAGEMENT OF NETWORKED INFORMATION

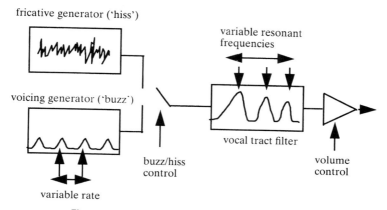

Figure 6.15 Elements of speech synthesis

The speech produced by systems like these is, to date, rather unnatural and can be difficult to understand over poor telephone connections, but research continues into more sophisticated models.

In the meantime, the most promising method is a halfway house between the fully synthetic method and the simpler tape recording method: fragments of speech, perhaps several hundred of them, are extracted from utterances made by individual speakers and used as building blocks for new utterances. The fragments do not correspond to 'pure' sounds (phonemes), such as 'k' or 'ee', nor do they correspond to complete words; instead they lie somewhere in between. perhaps groupings of three sounds – 'bre', 'ach' – and are selected pragmatically rather than on a rigorous theoretical basis. The synthesis approach is quite different from the complete recording case: the fragments can be used to generate words that have never been spoken by the originator, and the pitch and emphasis are generated automatically. What is preserved is a significant amount of the original character of the voice. In some cases it can be recognised as that of a particular accent or individual. So, it is technically possible to synthesise the voices of famous people, from past recordings, and make them say anything that is wanted; probably not sufficiently well to convince a jury but well enough to convey the 'essence' of Humphrey Bogart in an advertisement, for example.

Our best efforts do produce speech that is highly intelligible over a virtually infinite vocabulary. The speech can have quite natural inflections, for instance showing the difference between a question and a statement. Many synthesisers do sound as if they have originated from the voices of 'real people' and some can even offer a choice of speaker. However, all of them manifest unnatural distortions that do not occur with live talkers and would probably be recognised as produced by non-human sources.

They are, however, a fully competent way of providing easy to use interaction, even without further development. Just as with speech recognition, their method of working owes little to deep-level linguistic structure and much more to a pragmatic approach. Unlike recognition, the computational overhead to carry out the current algorithms is not great and synthesis can easily be achieved in real-time on desktop machines.

SYNTHESISING FROM TEXT

It is not sufficient just to take the raw text, on a letter by letter basis, or even in terms of groups of letters corresponding to our speech fragments, and use them to drive a speech synthesiser. Consider the text '*I live in a live way.*' Clearly, the word 'live' is pronounced in two very different ways. How do we determine the correct pronunciation in both cases? First the text is broken up into words and each word is checked against a database. Ignore, for the time being, words that are not found. Some words, e.g. 'in', will fit uniquely, in that they have only one pronunciation. Other words have different pronunciations depending on which part of speech they are; e.g. 'live' is an adjective and also a verb. The database will give the phonetic transcript for each and also the corresponding part of speech, together with the probability of the word occurring in each case, measured from bulk statistics of text, e.g. 'live' as a verb 55%, as an adjective 45%.

Using the words that are fixed parts of speech and other statistics, the synthesiser will then parse the sentence. For example, it could begin:

I = pronoun, therefore good probability a verb will follow
live = verb or adjective (so, from above, live is a verb)
in = preposition, so phrase will follow
a = article, therefore adjective or noun to follow
live = verb or adjective (so, in this case, an adjective)
way = noun

This is a very simplified example of what actually happens, but it explains the general principles. In some cases, the rules will give only statistical answers and the parser will have to combine the statistics in order to estimate the most likely 'part of speech' of the word.

So, the choice of pronunciation of the words has been calculated and they can, in principle, now be broken up into the basic sound fragments described above, and spoken out. (Words that are not recognised could be spelt out letter by letter, or spoken using a best guess based on letter statistics.) However, this process would result in a flat, monotonic rendition. Instead, the synthesiser looks at the output of the parsing process:

> pronoun   verb   preposition   article   adjective   noun

and, again according to the conventions extracted from large numbers of real examples, decides how it might be stressed:

unstressed   STRESSED   unstressed   unstressed   STRESSED   STRESSED

On this basis, the duration and loudness of the words can be modified. The synthesiser will also look at the parser output in order to construct a pitch pattern appropriate to the sentence structure. Once this is done, all the information is passed to a signal synthesiser and audio unit which generates the spoken text.

We have gone through this process in some detail to show how relatively 'unintelligent' this system is. Notice that it does not 'understand' the speech in any real sense. The individual words appear to the synthesiser only as strings of bits that share a set of statistics on their joint probability of occurrence, taken from analysis of large volumes of data. The words have no 'meaning', they only have probabilities.

But, nevertheless, it works remarkably well. The point to note is that arguments about the possibility of applying a deep level of artificial intelligence to speech synthesis are, to a large extent in many operationally desirable applications, beside the point. We can produce very worthwhile results already, by brute force and observation alone.

THE INTERPRETATION OF IMAGES

Earlier we said that text and speech do at least possess an underlying set of rules, a 'grammar', however difficult it is to determine. It is much more difficult to see how this can be the case with images. After all, language does contain elements that we may loosely call 'nouns' or 'verbs' (although linguistic experts use slightly different terms). In

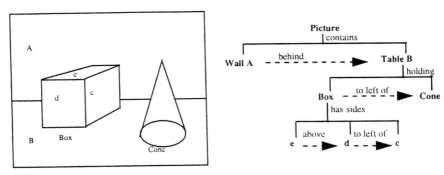

Figure 6.16 A scene grammar

fact, it is possible to write a grammatical description of a scene (Figure 6.16).

The grammar allows us to make deductions about objects in the scene. For instance, since side e ie above side d and side d is to the left of c, then we can deduce that e is also above c. Again, wall A is defined as being behind table B, which holds the box and the cone. This will imply that the wall is behind the box and the cone and, in consequence, the line separating the wall and the table will vanish from our sight whenever either of the two objects is in the way. One way to consider a grammar is as a set of constraints that limit the arbitrary layout of words in a text or of lines and colours in an image.

IMAGE RECOGNITION

The ability to manipulate images through the use of picture grammars is important, but how do we identify the images in the first place? We need to pick out the box, the cone, whatever, before we move them around. Some simple objects are easy to recognise: they are coloured or shaded distinctively from their backgrounds; they do not move, or when they do, they preserve a relatively constant shape (for example, a sphere). If automatic recognition is going to be generally useful, however, it must operate when some or all of these conditions are not met.

One technique that is widely used is 'landmarking': in many of the shapes we wish to recognise, there are usually a set of low-level features that can, in fact, be recognised more easily, because they maintain a constant shape when seen from different views, or have a significant contrast against their backgrounds. In the case of faces, for

INTELLIGENT MANAGEMENT OF NETWORKED INFORMATION

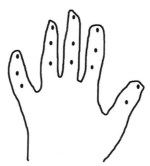

Figure 6.17 Landmark points on a hand

example, the eyes would provide an obvious landmark, both individually and as a pair. Locating both allows the system to calculate an approximate scale and orientation of the complete face. We can then begin to look in the right place for less prominent features – the nose, mouth, etc. – and the complete outline of the face. In Figure 6.17, we see a hand that has been 'landmarked' at the finger joints. We can define some 'grammar rules' for the landmarks. These are essentially a series of constraints on how the landmarks can move or transform relative to one another.

For instance, if the hand is formed into a pointing action and we have defined how the joints move, then an intelligent algorithm can deduce that the joints indicated in Figure 6.18 by the flat ellipses, and the rest of the fingers associated with them, will disappear from view on the far side of the fist, while the extended finger (whose joints are marked by x's) will remain visible.

This sort of information provides valuable clues to the recogniser; for instance, when examining successive time frames, it may notice that the landmarks on all fingers but the first are shrinking in towards the palm, as seen in a two-dimensional image. The recogniser's grammar

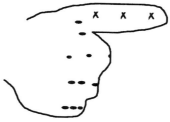

Figure 6.18 Movement of landmarks on pointing hand

could be organised to identify behaviour such as this as 'hand-like' behaviour and it would interpret this as the possible beginnings of a 'pointing gesture'. Next it would concentrate on establishing the extent and direction of the pointing finger: it would have valuable grammatical clues, knowing for instance that the landmarks of the pointing finger will be in a straight line and there will be certain average distances between them. This would allow it to select the approximate area to look for subtle changes in shade, colour or movement that might distinguish the fingertip from its background, ignoring extraneous locations where similar changes might fortuitously have occurred. Current research is actively looking at how such a system could act as the basis for remote control:

Instead of using a 'mouse' or a data-glove interface with a computer, we could imagine a system which recognised our gestures. It identifies a hand, perhaps because we move a hand into an area under a TV camera. It then recognises the fingers and, as they move, tracks them in the manner described above. Suppose the user wore a virtual reality mask: she could see in front of her a space filled with virtual objects and, superimposed on the image, her own hand (actually a reproduction of it, synthesised from the extracted image). The hand moves as she moves her real hand. Pointing at an object would open it up, allowing the contents to be displayed or someone's contact number to be 'dialled'. Practical examples of this are near – perhaps not sufficiently reliable to carry out delicate surgery or the operation of dangerous machinery, but it is quite feasible that natural interfaces of this type will become available within a decade, perhaps even less.

ARTIFICIAL NEURAL NETWORKS

The method described above relied upon the explicit identification of significant features in the image. A completely different approach is to use a system which implicitly derives a model from the data – one which is not necessarily easy to equate to a set of features. A good example of this type of recognition is the 'artificial neural network' technique, so called because it bears some resemblance to the way that, it is believed, the brain performs some of its intelligent functions.

Artificial neural networks consist of a very large number of simple, interconnected units that can perform logical functions. The ability of these networks to carry out their tasks comes about because of the high degree of connectivity between them, thus giving rise to another

INTELLIGENT MANAGEMENT OF NETWORKED INFORMATION

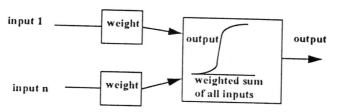

Figure 6.19 Single element in neural network

name for this class of techniques: 'connectionism'. Typically, a single element in a neural network behaves as described in Figure 6.19. Into the neural element are fed a number of the outputs of other neural elements, each via a weighting element which changes the strength of its effect. Each element may have a different 'weight' applied to it. The neural element then adds up all these weighted inputs and produces an output which is not their simple sum. In fact, it has a shape similar to the 'S' shape shown in the figure. This shape, which approximates to the way that a biological neurone in a brain behaves, reacts very little to small inputs and tends to saturate with big ones. This means that in the connectionist processes that comprise the network, the very small is ignored and the very large is not allowed to dominate too much.

The outputs of the elements of the neural network are connected into the inputs of other elements in the network. There are a number of configurations that have proved useful in practice. One common example is shown in Figure 6.20.

The neurones are stacked in three layers: an input layer which receives the data to be analysed, the output layer which gives the response, for example the recognition of one object among many, and a hidden layer.

In a typical image recognition example, we might have several hundred inputs to the input neurones corresponding to points on the image plane (pixels from a TV camera, for example) and 10 units in the output layer, each one corresponding to one object out of the 10 we wish to recognise. For successful recognition, we want this neurone to have a large output and all of the other output neurones to have very small outputs.

We start off by setting the weights with random values and expose the network to one of the test images. It is extremely unlikely that this will result in only one output node having a large value. However, it is possible to have a simple algorithm that, when told that the input

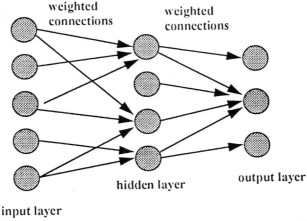

Figure 6.20 Typical connectivity of elements in neural network

corresponds to image n, say, then varies the weights in such a way as to drive the nth output to a high value and all the nine others to a small value. A second example of the same object class is applied to the input and the process of adjusting the weights carried out again. Other classes of objects re also applied and the weights adjusted in the direction of the required result. Gradually, under most circumstances, the network adapts so that, if we input a new example of image n, only the nth output will have a significant value. The network has 'learned' to classify its inputs.

In the example cited, the network was trained on the basis of adjusting the weights according to which class the object belonged to. It is even possible in some cases to carry out 'unsupervised training': the network spontaneously begins to separate out different clumps of data corresponding to the different classes of objects.

It sounds rather magical, but it is not. It relies upon there being separable properties inherent in the data from different object classes. These differences may be very difficult for human observers or classical statistical methods to discern, but they must exist. In fact, this is one of the problems with neural networks: they do detect differences between different classes of objects, but it is not always possible to see why. This means that one cannot in general use a neural network to extract 'features' that could be used by simpler or faster classification techniques; it also means that it is difficult to predict how a network that was successful on one class of problems will perform on another. Neural networks were greeted enthusiastically after an initial period of scepticism; some (though not all) of that scepticism has returned in recent years.

RECOGNITION OF FACES

The ability to recognise a person's face is something we all have and can use as a way of knowing how to greet someone as a friend or in a business context; it also acts as an effective means of preventing fraud or deception by impersonation. The introduction of videophones for remote banking and other transactions will clearly make it easier for humans to do so remotely, but there will be occasions where it would be cheaper or more convenient to identify someone's face automatically. The latest research in this area is yielding encouraging results: systems are under development that can uniquely recognise one face in several hundred.

One technique, known as 'eigenfaces', relies upon the idea of generating a 'recipe' for any one face, creating it from a mixture of different amounts of standard ingredients, or 'features', to use the terminology we have used elsewhere. Suppose we have a number of features, f_1 to f_N (where N will typically be 20–30). We then describe Ann as having a_1 units of f_1, a_2 units of f_2, and so on, to a_n units of f_n. Bob, on the other hand, is found to have b_1 units of f_1, b_2 units of f_2 and so on. When an 'unknown' face, X, is presented to the system, we measure how much of each feature it contains. We then compare this set of N numbers with our existing patterns, to find the nearest match, which corresponds to the most likely identity.

The question now arises, 'What are the features that are used as the basis for the recipe, and how are they extracted?'. It turns out that it is possible to apply statistical methods to the individual pixels that make up the images and automatically derive weighted collections of them that are very characteristic of 'faceness'. For instance, 'roundness' is a property that faces tend to have more of than do most objects. So, if we find that there is a significant pattern of changes lying in an approximately circular manner, within an image that is supposed to be a face, then we can reasonably infer that this circle is the boundary of the face. But some faces are 'rounder' than others and we can define a measurement of the roundness feature by the ratio of a over b, as shown in Figure 6.21.

So, our recognition strategy is first to locate the outline of the face, using our knowledge that it is approximately round, then estimate its

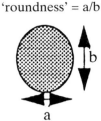

Figure 6.21 Measure of 'roundness' of a face

actual roundness (a/b) and search through our library of faces for the closest fit. As we said earlier, we actually choose to examine a number of features – for example amount of hair, or skin colouring, to name just two – and combine the results of the measurements.

Roundness, skin colouring and hairline are three features to which we can easily relate. Automatic extraction sometimes throws up features which are much less immediately equitable to human recognition, but the valuable thing about the processing is that it can be used to rank the underlying key features in terms of importance – that is, in terms of how significant they are for distinguishing between faces. The features that are produced (the weighted combination of groups of image points) are known as 'eigenvectors' and have the useful property that they also can be used to construct faces. A recipe for a face, 'take so much of this eigenvector and so much of that', can be used to synthesise faces that never existed, but which may resemble real individuals or groups of people. Combining this with speech synthesis, as described earlier, allows the creation of artificial personalities for use in games or as friendly, personalised interfaces.

REASONING ABOUT IMAGES

If we can recognise images, using the techniques described above, then we can reason about their behaviour. One good example of this is the work done at the University of Leeds in looking at the automatic detection of atypical behaviour in the interaction of objects – in particular, the interaction of a car-thief with a target car.

Suppose we have a surveillance TV camera trained on a car park and we process the images in such a manner that we can identify individuals and the cars in the park, using some of the techniques described earlier. (In Figure 6.22, the identification software has drawn

Figure 6.22 Car park surveillance – identifying people and cars (© 1998 IEEE)

a line around the outline of each car and around the outlines of two pedestrians.)

Next, having identified the people, we can also design a system that continues to track them as they walk around. This has to be reasonably intelligent and embody some rules that cope with the shape changes that naturally occur in the image. This means that we now can define a 'trajectory' for each person's movement around the car park (Figure 6.23).

Having identified the cars, the people and the movement of the latter, we need to consider how we represent the 'significance' of the movement within the scene, so that we can begin to reason about 'behaviour' and separate benign from criminal behaviour. Notice that we, as humans, are forming the value-judgments and creating the 'frame' within which the reasoning takes place. We need to reduce all the possible variations of normal and abnormal behaviour into a manageable set of measurements and decisions based on these measurements.

The strategy we adopt is to consider the behaviour of people in relation to the static cars. Figure 6.24 plots the distance between the person and the currently nearest car, as a function of time.

Figure 6.23 Car park surveillance – trajectories (© 1998 IEEE)

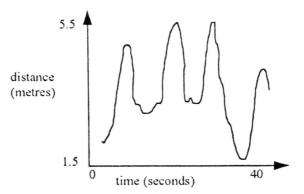

Figure 6.24 Distance to currently nearest car

It can be seen that this curve has some quite clear minima, corresponding to a person approaching a car. These minima can be used as 'anchor points' in the decision algorithm: we find the minima and then examine the behaviour close to them. We can measure the speed at which people approach the cars (simply, distance divided by time, measured close to the chosen minimum) and how close they actually get to the car. Someone walking across the car park will usually be some distance from most cars and travelling at a reasonable speed; even someone travelling along a line of cars will usually move relatively quickly. But someone closely approaching a number of cars and slowing or stopping beside them will manifest different statistics and can be treated as suspicious.

INTELLIGENT MANAGEMENT OF NETWORKED INFORMATION 219

MULTIMEDIA CLASSIFICATION

Much of the information available across the networks is going to be in the form of multimedia: text, sound and moving image. There will be far too much for us to search through it exhaustively. Consider, for example, all the footage of videotape that already exists in the libraries of television companies. When wideband networks and multimedia servers become affordable, all of this fund of information could go online, if we had a means to find our way around it. What we require is an 'information channel', like a continuous sound-track on a film but understandable to a machine. Unfortunately, virtually none of the material has been deliberately titled in this way. Suppose we could do something like this, automatically; what form would it take, and how could we use it?

We would hope there was a title for the video, at least. Perhaps this title contains a keyword which is referenced in our image library – say the word 'automobile', for instance. We may have a pattern recogniser trained up to locate objects that look a bit like automobiles. This recogniser could then be set to run through the video to identify instances where there was a reasonable fit to an automobile shape. The video is also likely to have a sound-track. This could be checked by a speech recogniser for instances of the word 'automobile', or synonyms, such as 'car', 'limo', etc.

Whereas books have words, sentences, headings, chapters and so on, the deliberate and obvious segmentation of video material largely died out with the disappearance of the caption-cards in the silent movies. This presents some problems when trying to find the boundaries for our intelligent searching. However, there still exist some more subtle syntactical/semantic clues within the material.

Camera operators convey meaning in a number of ways; a most simple and obvious example is a static frame. If the image does not change for an appreciable time, it is reasonable to expect that a single subject or topic is being presented. If we have already picked up title or speech clues that the subject material is 'landscape', for instance, we can be reasonably confident that a static frame implies a single landscape, rather than a succession. A 'fade' is just as obvious a clue: there is a transition in time, place or theme, before and after it. Slightly more complex, but still essentially 'unintelligent', analyses of the movement of images in a scene can be used to create intelligent guesses regarding what is going on.

If, as in Figure 6.25, there is a mass outward migration of points

Figure 6.25 Zooming-in

Figure 6.26 Panning

between one image and the next, then we are being given a close-up, corresponding to the semantic focusing in on the central subject. An overall drift of picture points in a parallel direction, resulting from a panning motion of the camera, represents a changing theme (Figure 6.26).

Combining these guesses with the object detection techniques and the speech recognition outputs begins to give us a way to analyse scenes for content and context. The fully automatic classification of video material is still a long way away, but techniques like these can at least act as a pre-processing filter to human operators.

A SUMMARY OF RECOGNITION STRATEGY AND ITS RESULTS

We have covered, in a very short text, a large and complex field, in order to give a flavour and basic understanding of the techniques used and an idea of their strengths and limitations. We can summarise quite succinctly: recognition schemes vary in the extent to which they impose a 'semantic and syntactic framework'; that is, how tightly they define the *features* that have a separate identify within the mass of raw data, the *rules* which describe their interactions and the *knowledge framework* within which the features and rules sit. Where features, rules and frameworks are sparse or absent, for instance in the case of speech recognition using only acoustic measurements, accuracy is usually rather poor. With tighter semantics and syntax, better results can be achieved, but sometimes with spectacular errors, and almost always a human being has first had to instruct the system as to what is to be considered meaningful.

The performance is strongly dependent on the nature and extent of the field of discourse. A limited number of voices or faces are much easier to separate out, as is a tightly constrained dialogue. There does not appear to be a single 'optimum recognition technique': mixed strategies using multiple data sets (voice, text, image) appear to give the best results, although it is extremely difficult to compare one strategy with another.

Trying to answer the question 'How far can we go in the artificial understanding and processing of information?' is therefore impossible in simple terms. Any successful strategy will be a mixture of genuine artificial intelligence and practical low cunning (for example, dialogue design). We are a long way from designing machines that can approach the performance of humans in any complex task. Fortunately, as we discuss further in the final chapter, there are many tasks carried out by humans that are probably not as complex as we believe.

7

Technologies for a Wicked World – Trusted and Trusting Networks

SCENARIOS OF TRUST

Today, about 5% of goods, cars excluded, are purchased remotely by mail order or telephone, mainly by the latter. At least one respected forecasting company predicts that intercompany trading over the Internet will grow fortyfold [from 1997] . . . to $8 billion by the year 2002. If these figures are to be believed, we are going to have to build networks that protect both our secrets and our money. Consider the scenarios described in the following paragraphs.

Securing Information   Business A and Business B have gone into a trading partnership. A makes components for B and B incorporates them into a product and markets it. A hopes the strategic partnership will save him from having to spend time looking for multiple outlets for his components; B hopes to guarantee a source of quality supply without the cost of continually assessing suppliers. They both hope that the partnership will reduce time-to-market because they will not need to renegotiate formal contracts. But how do they know that they can trust each other? One thing they can do is to share their databases; A lets B have access to the current stock-holding position, while B reveals the number of orders in the pipeline. These are very important and sensitive pieces of business information and it is vital that they are kept secret. Neither A nor B is willing or able to maintain cumbersome security regimes, and both are very likely to lose or at least compromise any equipment they use for security purposes. Nor do they live in a trading vacuum: A also trades with C, D and E, and B with X, Y and Z; they share data with them and are always on the look-out for new business. They need ways not just of protecting data as a whole, but of partitioning it into domains of trust between themselves and a closed set of partners, which may change rapidly.

Identifying Who's Who   Medical records are among the most sensitive pieces of information, and yet, in an emergency, they may be the most urgently and easily required. How would a database at the other end of the world know whether to release the information to someone purporting to be treating a disaster victim, if that person had no way of communicating other than by telephone? How reliably can we trust the identity of a person, by their voice or by their signature, for instance?

Suppose a local doctor wants to send your medical records to me, a specialist consultant of international fame, but unfortunately at the other side of the world and not known to her personally. The records comprise megabytes of data and urgency is of the utmost importance. Perhaps I wish to examine you remotely or even carry out invasive tests that could save your life. How can the information be quickly coded up, using a publicly available security key that can only be 'unlocked' at the other end by me? How can your doctor arrange that it is only I that can operate the tele-surgery equipment?

Secure Payment   Buying things with money is more complicated than it looks. How do we know that the money is genuine? How do we prove that we paid and how does the seller know that our credit will be honoured? If we buy things in dangerous places – in an open market, or on the Internet – how do we stop someone stealing our wallet or using our credit card for fraudulent transactions? Problems such as these bedevilled trading long before it became electronic and a number of solutions – signatures, PINs, photographic pass cards, etc.– have been developed, but each one has a cost attached to it. Carrying cash is a simple solution, but it has its own problems. What is required for electronic trading is a range of strategies that provide security at an affordable overhead cost on each transaction.

ISSUES OF TRUST

In trying to understand issues of security, we need to remind ourselves that security itself is not the prime aim; the real task is the construction of systems that generate and deserve trust. Much of life is about trust: we trust our friends and colleagues to cooperate and to tell us the truth; we trust doctors and lawyers to keep the secrets that we give them. Business is built on trust: betrayal of commercial secrets can lead to bankruptcy. Financial dealings, from simple cash purchases to complex interbank transfers, all rely on confidentiality and enforceable penalties against fraud and defaulting. If we are really to see a major increase in networked organisations and electronic commerce, we need to be able to provide networks that protect against breaches of trust at least as effectively as face-to-face transactions. Trust is complex and rather difficult to define. That is why we tend to talk instead about 'security', but we must remember that it is fundamentally trust that we are trying to preserve.

DEFINING 'SECURITY'

'Security' used to be about ways of allowing allies to exchange secrets without an enemy being able to understand what was being said or being able to inject misleading data, but now it embraces a wide field of activities:

- Keeping secrets: this is probably closest to the original sense. Information such as sales figures, medical records, bank accounts and passwords are worth money or can cause damage if they get into the wrong hands. How do we pass secrets between us, without someone else finding them out or modifying them, and how do we do so in a way that will not go wrong in practice and is not going to be so cumbersome that it will inhibit our easy communication?
- But how do I know that it is you that is really there and not another person? For instance, if I run a bank, how do I know whether to allow someone who *says* that he is you, to withdraw funds from your account?
- And how can I guarantee that a request, a purchase order for instance, came from you, even if you subsequently denied it?
- What about electronic money, in general? How do I know it's 'real'? How do I transmit it securely?

Table 7.1 Terms for security features

Common terminology	Description
Authentication	Are you who you say you are?
Access control	Are you allowed to have access to this?
Confidentiality	Is the information restricted to certain people?
Privacy	Can anyone observe what's going on?
Integrity	Can the transfer of information be corrupted?
Auditability	Do we have records of what has taken place?

A number of standard terms are used to describe these facets of security (Table 7.1).

BASIC PRINCIPLES OF SECURE COMMUNICATION

Suppose two parties, A and B, want to communicate something valuable or confidential across a network (Figure 7.1). Then, obviously, they must secure each part (i and ii, in this case) of the network. They will also have to secure the network nodes and their terminals, 1, 2, 3. One way to do this is to make all parts of the link impossible to intercept, and we shall discuss how this might be done in an all-optical network.

An alternative might be to employ end-to-end encryption of the information. To do this, they will also need some way of managing the transfer of the keys that are necessary for encrypting and decrypting the information. So we have to consider the development of a key

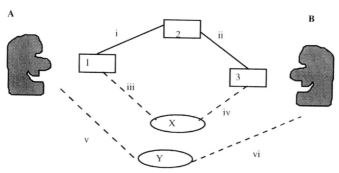

Figure 7.1 Basic principles of secure communication

management service X, and links iii and iv, for transferring the keys. We shall see that, under certain circumstances, the physical links i and ii can be used to convey the key information also, although they are insecure.

Even supposing the link is secure, such that no one can eavesdrop on the information carried over it, there is still another security issue. How do the parties at either end know that they are communicating with the right person? We need some way of knowing that A is A and B is B, and a mechanism Y, v, vi, of agreeing that this is so.

Finally, before going on to look at techniques, we need to consider the practical aspects of security systems.

Firstly, no security system is going to be effective if it is too difficult to operate by the people who are intended to use it. There are well-recorded cases of intercepted military wireless messages where operators ask each other, on unencrypted voice channels, for passwords and encryption keys, thereby rendering null and void the highly powerful encryption equipment they are about to use. There are also cases where passwords were not changed frequently enough because the equipment was too difficult to set up.

Secondly, equipment can go wrong, either by accident or through deliberate sabotage by a disgruntled employee. It can also be stolen. As a good general rule, it should be assumed that your 'crypto' equipment is understood by your enemies. It is only the key settings that you need keep secret.

Thirdly, you need a security policy, not just a technology, whereby you assess risk of attack and consequences of it being successful, so that effort and inconvenience can be balanced against them and the users can accept the need for secure behaviour.

Finally, it should be remembered that security is a task for the ingenious expert; amateur or pedestrian efforts can be worse than useless. To take a simple example, we all 'know' that dial-back modems provide protection against someone trying to pretend to be who they are not, but see the frame, 'Defeating the dial-back'. In the

DEFEATING THE DIAL-BACK

A well-known technique, intended to foil an attempt by a hacker to log in as a legitimate user, is to arrange for the computer service to dial back to the legitimate user's telephone. It is widely believed that this makes it impossible for a hacker to gain access. But consider the following diagram:

TECHNOLOGIES FOR A WICKED WORLD

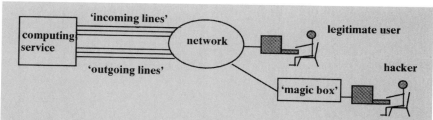

A number of telephone lines connect between the computer and the network. Some of these have been designated 'incoming' lines by the computer company, who publish their numbers. Other lines, 'outgoing' lines, are used by the computer to dial back. Their numbers are not published, but it is often easy to work out what they are from the known numbers of the incoming lines, or by getting inside information. Although they are called 'outgoing', in many cases they are normal telephone lines, without any incoming call-barring. This means that *they can accept incoming calls*. Equipped with what the literature would probably call a 'magic box', a hacker makes a call to one of the 'outgoing' lines. Assuming the line isn't busy, she (or rather, her magic box) hears ringing, generated of course at the telephone exchange. The hacker holds on. Meanwhile, the legitimate user calls in, logs on, maybe even gives a password. 'Please clear down, I shall call you back', says the computer. The user puts her phone down, releasing the call. The computer now selects an 'outgoing' line. Suppose it selects the line currently accessed by the hacker. As soon as the computer connects to the line, it has, unbeknown to itself, accepted a call from the hacker. The hacker's magic box detects the fact that ringing has ceased and immediately begins to generate a sound on the line that sounds like dial tone. The computer is waiting to hear this 'dial tone'. As soon as it does, it (quite pointlessly) dials out the first digit of the user's phone number. This is detected by the box, which, mimicking a real call setup, removes the dial tone. The computer continues with the futile exercise of dialling. Once it has finished, the hacker responds with whatever protocol is required to let the computer know that the 'user' is back on line. The hacker is in! Probably beyond password level too.

There are ways round this. Probably little harm is done in disclosing a technique that has been known for so long, but there are organisations that should know better, who have fallen for it in the past.

technological solutions that follow, it is possible to see how some of these practical issues have been considered; for example, the success of public key encryption lies in its convenience rather than its cryptographic strength.

SECURING THE LINK

Before looking in detail at methods for protecting information on insecure links, we briefly touch on methods for making the link impossible to intercept, which are usually employed in special circumstances only.

Until very recently, it was true to say that it was never possible in principle to make a link entirely secure. There were many links which were practically secure: it is unlikely that, in peacetime, anyone has gone to the bother of trying to intercept a transatlantic submarine cable, which would involve a great degree of expense and the problem of operating in deep water to cut into a cable powered by several hundred volts, and even land-based coaxial cables present significant problems, but nothing could be guaranteed. It has been said that conventional optical fibre cables are resistant to interception. This is true, to a degree. Because of their fragility, when stripped of their protective coating, they are difficult to work with in secrecy and in the hidden locations where they might be attacked, but attack is possible. The same principle that is used legitimately to construct an optical splitter can be used to make an optical tap. A variant of this (see Figure 7.2) uses 'microbending' to create small-scale periodic bends in the fibre that allow light to leak out into a surrounding optical material, for example a clear epoxy resin, into which is placed a diode receiver.

One way to defeat this method of attack is to make use of the fact that an optical fibre consists of two concentric glasses, a central 'core' surrounded by a 'cladding' (Figure 3.4), and to inject a spoiling signal into the cladding (Figure 7.3). Anyone trying to intercept the transmission by leaking out light by the techniques we have mentioned,

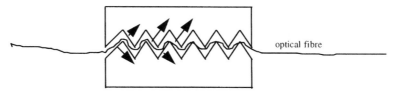

Figure 7.2 An optical 'tap'

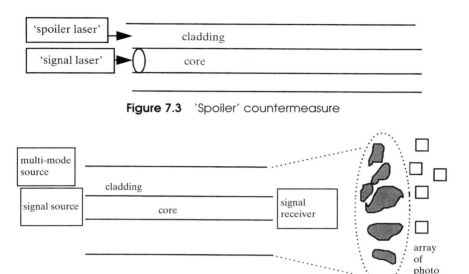

Figure 7.3 'Spoiler' countermeasure

Figure 7.4 Multimode security signal

would obtain a signal which was heavily masked by the spoiler and would probably not be understandable.

A variant of this technique is shown in Figure 7.4. We generate a spoiler signal which is sensitive to interception in the sense that its properties are modified in a way that can be noticed by the legitimate receiver at the far end: the optical cladding is relatively large, of the order of at least several tens of microns, and can therefore support a number of 'modes' of light propagation. This is roughly equivalent to saying that a number of rays of light can pass along the cladding. A multimodal laser injects a signal into the cladding at the transmit end. At the receiving end, the interference between these various optical modes creates a 'speckle pattern', a set of dark and bright patches which depend on the exact mix of modes, their amplitudes and their relative optical path through the cladding. Any alteration to any these, as would be brought about by an optical 'tap', will cause an alteration to the speckle pattern, which can be detected by using an array of photodiode receivers.

QUANTUM CRYPTOGRAPHY

It is theoretically possible to get round these security measures by very careful design of the interception equipment, but a solution has recently emerged that claims to offer undefeatable protection against unauthorised interception. The multimodal system described above works on the principle of generating an alarm if anyone taps off a portion of the signal, but it is possible to see that, in principal at least, the attacker could reintroduce the leaked signal, to exactly compensate for what had been removed. Suppose, however, we could use some phenomenon that fundamentally disallowed such a correction to be made. Quantum cryptography does just this by exploiting the 'Uncertainty Principle' of quantum physics, an experimentally verified theory that demonstrates the impossibility of carrying out the simultaneous, exact measurement of certain pairs of physical values. The act of making a precise measurement of one of those values, for example the momentum of a particle, such as a photon of light, introduces a disturbance into the other value (the position, say), and the uncertainty created is proportional to the exactitude of the other measurement. This implies that any attempt by an eavesdropper to make a measurement of the messages fundamentally and irreversibly introduces some corruption into the message, which can be detected if the sender and the legitimate receiver 'compare notes' on a sent and received test message.

Undoubtedly, quantum cryptography promises to be significant for applications requiring ultra-high security, but it is still experimental and very difficult to implement, and may have a measure of overkill for most requirements.

FIREWALLS

Protecting date from attack is not the only problem; the telecommunications and computer systems themselves are vulnerable to attack and the consequences can be very serious. Imagine, for example, what would happen if someone were able to hack into the main system of a major bank and destroy all the customer records; less hypothetically, recently a large fraction of the world's e-mail was returned, marked 'unable to deliver', because of a fault in the master database of the e-mail addressing system. This was caused by operator error, but, in principle, is the sort of thing an outside hacker or a disgruntled employee might try to achieve.

In planning a defence against an attack, the best thing to do is to rely upon the tried and tested security principle of cutting down the number of access points and then patrolling these points with rigour.

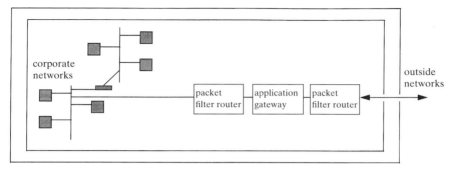

Figure 7.5 Firewall design

This is the principle of the 'firewall' (Figure 7.5). We can have as many and as complex internal networks as we like, but they all connect to the outside world through one pair of filters, one for checking outgoing packets and one for checking the packets coming in. The filters are conventional routers enhanced to include the ability to inspect the addresses of every individual packet. The systems administrator (who must be someone who can be trusted – an issue in itself) sets up a table for each filter to check against, to see whether it is OK to forward the packet. If forbidden addresses turn up, they are rejected and the administration alerted. By means of the application gateway, filtering can also be carried out, not just at the packet level, but also on the specific contents and purposes of the data passing through. For instance, it is possible to set up a free transit for e-mail between the corporate network and the external world, but to block off every other application.

ENCRYPTION

In most cases we cannot rely upon the link between secure from end to end: with radio paths or electronic amplifiers there exist obvious interception points; even in all-optical links, we have seen that tapping is a possibility, and on simple systems, such as local loop passive optical networks, signals destined for a number of recipients are narrow-cast on a single fibre. Thus, we require some way to encrypt them, so that only the legitimate destination can make sense of them.

The process of encrypting messages antedates electronic, or even electromechanical, technology, but with their arrival, it has explosively developed in both theory and practice. The principal components are

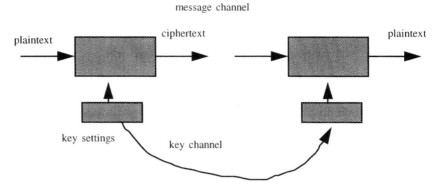

Figure 7.6 Components of encryption and decryption

shown in Figure 7.6. The message to be protected, or 'plaintext', is fed into an encryption machine, sometimes known as a 'crypto'. The crypto acts on the message according to a set of instructions known as the 'key' or 'key settings', that are changed on a regular basis, to produce an output, the 'ciphertext', which is intended to be meaningless to an attacker. The legitimate receiver of the message is equipped with a compatible decryption equipment and key settings, by the use of which the original message can be retrieved. Notice that two transmission channels are shown: one for the message and one for the key. Perhaps the major issue of encryption is 'key management and distribution': how the sender and receiver can achieve an advantage over the attacker. The crypto is usually considered to be something, a piece of hardware or software, that can be acquired by the attacker. Thus, its principle of operation is known. On the other hand, its precise mode of operation at any specific time is determined by the key. Thus the key must be kept secret.

A significant issue is the partitioning of the encryption between the crypto and its key settings. One of the simplest cryptographic techniques possible is also the most secure, the 'one-time pad' (see frame).

The 'key' for the method consists of the numbers on the pad; the 'crypto' is simply the operation of adding the key to the message. Although it is the most secure system possible, the one-time pad has a serious problem: the size of the pad has to be the same size as the sum of all the messages that you want to send. Reusing the pad exposes a cryptographic weakness, as explained in the frame. This may not be a problem where the messages we want to send are short and infrequent, and one-time pads are still, reportedly, found in the possession

ONE-TIME PAD ENCRYPTION

The letters of the message to be encrypted are first turned to numbers: A = 1, B = 2 and so on.

S	E	C	R	E	T
19	5	3	18	5	20

The enciphering agent has a set of pages containing lists of numbers, within the range 1–26, that have been generated at random: 21, 3, 13, 19, 5, 7, etc. These are added in turn to the numbers corresponding to the plaintext. If the sum exceeds 26, then 26 is subtracted. This gives the ciphertext:

	19	5	3	18	5	20
plus	21	3	13	19	5	7
gives ciphertext	14	8	16	11	10	1

If the numbers have been generated truly at random, then there is no way that any interceptor can work back to the plaintext. On the other hand, if the legitimate receiver has a copy of the random numbers and knows where they start, it is simple to subtract them from the ciphertext (and then add 26 if the result is negative) to get back to the plaintext:

ciphertext	14	8	16	11	10	1
minus	21	3	13	19	5	7
gives plaintext	19	5	3	18	5	20
	S	E	C	R	E	T

There is weakness from using the pad over and over again. One line of attack is as follows. Suppose the first message is $a_1\ b_1\ c_1\ \ldots$, the second is $a_2\ b_2\ c_2\ \ldots$, the third is $a_3\ b_3\ c_3\ \ldots$ and so on, for n messages. Just look at the first letter a, of each message. In each case, this is added to the same number, k, from the one-time pad:

$$a_1 + k$$
$$a_2 + k$$
$$a_3 + k$$

Add them up: T (approximately) $= n.m + n.k$ where m is the average value of a over all messages. This depends on the statistics of the language of the message. Therefore, k (approximately) $= (T - n.m)/n$. Thus we can find the values of the pad and therefore decode the messages.

of spies who wish to communicate messages of the 'meet me at 12.00 at the usual place' variety. However, if we want to encrypt a TV programme so that only paid-up viewers can see it, we would require one-time pads providing several megabits per second. The controller of a spy can, presumably, arrange to meet frequently enough to hand over new one-time pads, but it is difficult to imagine, with today's or even tomorrow's technology, how we could arrange a key distribution channel for every TV subscriber.

Clearly, we need a more practical solution. This is achieved by sacrificing perfect security in favour of a much shorter key and a more complex crypto. There are a large number of ways of achieving this, some based on the principle of generating numbers that appear random, which are then used essentially in the same manner as the keys of the one-time pad, others based on shuffling the data in the message in an order based on an apparently random key. A very simple example of the former (which is cryptographically very poor) is the 'linear-feedback shift register', shown in Figure 7.7.

A shift register allows bit-patterns to move sequentially from left to right. Some of these bits are copied via the closed switches into the 'OR' function device. This performs a series of 'half-adding' operations (0 + 0 = 1, 0 + 1 = 1, 1 + 1 = 0) on the data, such that an even number of 1s will give answer 0 and an odd number will give 1. The answer is then OR'd with the first bit of the plaintext, thus encrypting it. The answer is also 'fed back' into the left-hand cell of the shift register after all the other bits in it have moved one place to the right. This is an example of a 'stream cipher': a stream of pseudo-random bits is generated by the crypto and added bit by bit to successive bits of the plain text.

It is possible to show that legitimate receivers of the data can

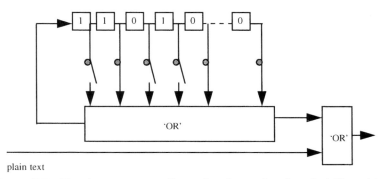

Figure 7.7 Keystream generation using linear feedback shift register

decrypt it with a similar device, provided they know which switches are closed and the start sequence (110100, in this case). The 'crypto' consists of the shift register and the function generator (an OR function, in this case). It has to be assumed that the attacker can have at least some knowledge of the number of stages in the shift register and the nature of the function generator. The only security that can be assumed is that contained in the key. There are two components to the key, in this device: the start sequence and the switch settings.

Practical systems use functions other than a single shift register with feedback and a simple OR function. While the single shift register provides some security when the attacker has no knowledge of the contents of the message, it is extremely weak when some of the text is known or can be guessed. (For example, many letters begin 'Dear Sir', contain the date, or end equally conventionally.) In this case, it can be shown that the system can always be broken, given less than $10N$ successive bits of plaintext, where N is the number of stages in the shift register.

There are two important points to be drawn from this: firstly, crypto systems must consider practicalities, as well as theory, and in particular the possibility of the attacker starting with more than zero knowledge. Secondly, what we must be trying to achieve is presenting a degree of difficulty to the attacker that increases faster than any increase in the complexity of the crypto. This is not true in the case of the shift register, where the complexity of the system and its susceptibility to attack both go up 'linearly', that is, with N. We would like a system whose complexity could be increased by N, with a resulting increase in difficulty of attack going up by (at least) N^M, where M is some large number.

Some practical alternatives offering better security include cascades of shift registers, whose modes of operations interact. For instance, we can use the output of one register to control the clock controlling the following one; that is, make it move on more than one shift at a time.

BLOCK CIPHERS

An alternative to stream ciphers are 'block ciphers' which operate on a block of data a fixed number of bits long to produce an equivalent, encrypted output. Rather than generate a separate stream of pseudo-random bits and combine them with the plaintext, block ciphers often carry out 'scrambling' operations on the plaintext data alone, reordering the bits in the block and combining them with other bits, for example using an OR function. For a very simple example, see the frame on how to cheat at cards.

> ### BLOCK CYPHERS: HOW TO CHEAT AT CARDS
>
> In the game of bridge, all 52 cards are dealt out, one at a time, to four players in succession. Let's say player N (for 'North'; this is the way hands are conventionally named) has cards 1–13, E (East) has 14–26, S (South) 27–39 and W (West) 40–52. In a 'duplicate' competition, several teams play the same deal, to see who can make the best score. The organisers often deal the hands before the event. One of them wants to help you cheat!
>
> She takes another pack of cards and arranges them in the same order as the deal; that is, the top card is 1, followed by 2 and so on. She then cuts the deck exactly in half (26 each half) and performs a 'perfect shuffle': this involves interleaving the two halves of the deck so that they merge one card from one half with one card from the other. If we look at the deck we see that they are now ordered 1, 27, 2, 28, 3, 29, and so on. Suppose she does the same thing again. She then leaves the cards in a public place, perhaps with two pencils. Anyone not in the know who examines the cards will see an apparently random order, which will not lead them to suspect they contain information on the deals to be played. You, on the other hand, 'idly' deal the cards alternately into two piles and, when they are all dealt, place the first pile on top of the second and repeat the process, because the two pencils told you she had executed two perfect shuffles. The cards are now in order 1, 2, 3, and so on. You have been given an advance look at the game that will be played!
>
> The 'perfect' shuffle is an example of a block enciphering algorithm or 'crypto'; the key is '2', equal to the number of times it is to be performed.

Several, more practical, examples are based on 'Feistel ciphers' which repeatedly pass the text through the same scrambling function ('F' in Figure 7.8).

A block of plaintext data is split in half. One half of the data is operated on by the function, F, which carries out a reordering (shuffling) of the bits in the block, according to a 'subkey;' (SK1). The output of

TECHNOLOGIES FOR A WICKED WORLD

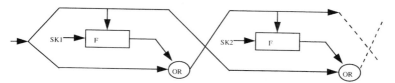

Figure 7.8 Two stages of a Feistel cipher

the reordering is then OR'd with the untreated half. The halves are then swapped over and the same reordering function is now applied to the other half. Note that the reordering function is still the same, but its effect on the data may be different because it is controlled by a new subkey (SK2). The process can be repeated over and over again. Feistel ciphers are convenient to decrypt, because the process of encryption is simply reversed, applying the subkeys in reverse order. The reordering functions are chosen such that their detailed operation on the data is difficult to decipher without knowing the subkeys. Thus the reordering functions can be made public.

THE DATA ENCRYPTION STANDARD (DES)

One published and extensively used example of a Feistel encryption system is the Data Encryption Standard, or 'DES'. This uses a key typically at least 56 bits long. This provides nearly 10^{17} possibilities that an attacker would have to try in a brute force attack. A well-programmed, general-purpose computer in today's technology would take 100 years or so to test every possibility. Doubling the key length to 128 bits increases the encryption time by the same order, but makes the brute fore attack 10^{20} times longer to do. This is impressively difficult but probably not the correct way to go about an attack. Conventional computers, which operate serially on a problem, one step after another, are not the best devices for code breaking.

One proposal envisages a very large population of patriotic homeworkers and a level of not very expensive, but widespread, technology. The emergence of satellite broadcasting in China has suggested a particular technique and the name 'the Chinese lottery', as shown in Figure 7.9.

The attacker has intercepted the link and transmits the signal on a satellite broadcasting channel into many millions of homes, each of which is equipped with a TV set and a satellite decoder, in the form of a set-top box. The latter is, essentially, a modified computer, and

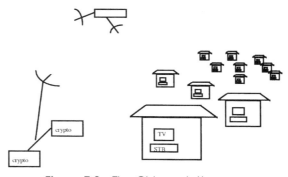

Figure 7.9 The Chinese lottery

could simply be further modified to allow the crypto algorithm to be included. Keys selected at random by each household are tested out to see if a meaningful message appears on the TV screen. One of the many million attempts will soon be successful. It has turned out that one of the earliest successful attacks on an encrypted message, using this technique, was achieved not through any great act of patriotic solidarity but rather by a volunteer force of 14 000 Internet users, one of whom managed to find the 56-bit key. A previous success using 3500 personal computers broke a 48-bit key in 13 days. The serious implications are that attacks based on heavily parallel computing will always pose a threat to any encryption techniques. It has been estimated that special-purpose machines, costing around $1 million, that carry out a large number of parallel attacks on a 56-bit DES could achieve a success within several hours. This is well within the means of government security departments of many nations (and probably of several global criminal syndicates). There is also a continuing conspiracy theory that DES contains a weakness that was intentionally introduced by the National Security Agency, but despite there being many attempts by leading experts opposed to government 'interference', no successes have been published, and the story must be treated with some scepticism.

Other attacks have been proposed that use 'plaintext' approaches, where the attacker manages either to get the sender to encode selected messages chosen by the attacker, or simply uses messages whose plaintext is known. One example of the latter successfully recovered the key in 50 days using 12 workstations, but this required 2^{43} known plaintexts, and thus is probably not practical.

In any case, multiple applications of DES are now deployed, together with long keys, and, in a world where nothing is certain, it is probably true that DES is more secure than is the practical management of its keys. We now look at how some of the key management problems can be alleviated.

PUBLIC KEY SYSTEMS

The systems we have looked at above have all been concerned to achieve a high degree of security with a manageable arrangement for key handling. But they all share the common problem that there still has to be a secure channel for delivering the secret keys. At first sight, this seems an inevitable requirement; if you find out the key I use to encrypt a message, then surely it is easy for you to decrypt it? This is true if the key for decrypting a message is either identical to, or easily derivable from, the encryption key and the decryption method is a simple reversal of the encryption. But suppose this is not so. In this case, it turns out that there are a range of techniques which open up surprising possibility: *the ability to publish openly an encryption key for anyone to use, that does not make it easy for anyone, other than the issuer of the key, to decrypt any messages encoded with this key.*

Public key encryption systems rely on the fact that some mathematical operations are easy to perform in one direction but difficult in another. One example of this is forming the square of a number – an easy operation of multiplying the number by itself – compared with being given the result and asked to find the square root. The number of steps required to do so is much greater than for the squaring operation. Another, more relevant, operation is that of 'factorising products of prime numbers' (see frame). It is easy to multiply two prime numbers together, but difficult to reverse the process.

PRIME NUMBERS

Prime numbers are whole numbers that cannot be 'factorised' into the product of a number of smaller whole numbers: 1, 2, 3, 5 and 7 are primes, but not 4 (= 2 × 2), 6 (= 2 × 3) or 9 (= 3 × 3). It is known that there are an infinite number of primes, but no formula is known for proving, in general, that a number is or is not a prime, that is significantly quicker than trial and error. The time taken to do this increases very rapidly with the size of the number A similar problem is that of 'factorising the product of two prime numbers', i.e. finding p and q, given only p times q. There are a very large number of prime numbers: about 10^{150} of them with 512 bits or less.

THE RSA ENCRYPTION ALGORITHM

The algorithm is based on selecting two large prime numbers, p and q, whose identity must be kept secret. Let $n = p \times q$ and choose a number e, which is less than n; e need not be a prime number but it must not have any factors in common with either (p − 1) or (q −1)

We now need to find another number, d, such that $(e \times d - 1)$ is divisible by (p − 1(q −1) without a remainder.

Thus we have three numbers: d, e, n. We make e and n public (they are the 'public keys'). We keep d (the 'private key') secret.

Suppose Alice wants to send us a message, m. She looks up our value of e in a public list. She then raises m to the power e (that is, she multiplies m by itself, e times. If m was '345' and e was '7', she would calculate $345 \times 345 \times 345 \times 345 \times 345 \times 345 \times 345$).

She then finds n from the public list and repeatedly divides it into the previous calculation until only a remainder R is left, (e.g. suppose m to the power $e = 10n + R$; in mathematical notation, this is known as finding $R = m^e$ mod n). R is the ciphertext that Alice sends to us over an open channel. It can be shown that any technique that allows m to be found from R and knowledge of n and e is computationally difficult, equivalent to trying to factorise the product of two primes.

We take the ciphertext R and raise it to the power d (the secret key). We then find the remainder after repeated division by n (i.e. we have calculated R^d mod n). Surprisingly, this turns out to e m, the plaintext massage.

A typical public key value, coded as an alphanumeric string, would look something like this:

iQEVAwUBNOW/58UCGwxmWcHhAQG+Mgf/RRBnjaC5ir
0QKuD8+tgNhhvdoi2ajZRkKgaBzsd0PDRy/3W2UJTuIR/pz
MQBRKp0pstRVkTO74edyIHEWwbUcIvxCmQaCx69NC2sDR
iNDBDyGdfLHq4k7EoEE+Z5KY1+bcXLJuIPHdBKpNbDZ9I
Prs5z2S0gyc1IFPZkj+EjO8oNXXA9H9nqzhF5aHvLv5XZrxvb0
8iLAM16iepBQz7qdh617CPm7tu7A8VD9G+46kGXN4snjOjR
UXgKnoQGMTs1TBdMWuse6qjipSMLxXVQXCvAOoD8VScg
31I3EQB7mjX+P2nVchVYVUBfpxgYJsvtw5zqBBd3f1tpeLCeI2
9Sg7/Fzwi

TECHNOLOGIES FOR A WICKED WORLD   241

THE RSA PUBLIC-KEY CRYPTOSYSTEM

The difference in difficulty between the multiplication of prime numbers and the reverse process is the basis for one of the most successful public-key systems, 'RSA', named after its inventors, Ron Rivest, Adi Shamir and Leonard Adleman. A full proof of the technique requires some complex mathematics, but the principle is simple (see frame). Two large prime numbers of approximately equal size are selected and multiplied together to produce a number several hundred bits long. This is called the 'modulus'. Two further numbers are calculated from the two primes. These are called the 'public' and the 'private' keys.

The owner must keep the choice of the two primes and the private key secret from anyone else, *but can make the modulus and the public key openly available.* Thus there is no need for the complexity of setting up a secure channel for passing our secret keys.

Anyone wishing to send a message uses the simple algorithm described in the frame to encrypt the message, operating on it using the modulus and the public key. Once the message has been encrypted, the task of decrypting it is equivalent to trying to factorise n, and this is known to be 'difficult' unless one has access to the private key. Knowledge of this private key allows the legitimate recipient to decrypt the message using a similar algorithm to the encryption, except that the private key is employed instead of the public key.

VULNERABILITY

How difficult is 'difficult'? The answer to this must be given with some caution, as the general mathematical theory of factoring and its application to encryption are both very active fields. Some sets of prime numbers are known to be weaker than others and there are possibilities for plaintext attacks, but current opinion is that RSA and similar systems are highly resistant.

Recently, claims have been made that rather exotic devices, 'quantum computers', may significantly upset the balance between ease of encryption and difficulty in decryption, by providing a means to carry out massively parallel processing on a single chip. Successful methods of attack need not, in any case, rely on technological or mathematical breakthroughs. Issues of national security will never make it possible for us to assess how many ciphers are broken by purely cryptographic

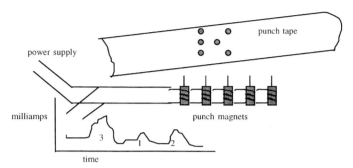

Figure 7.10 Back-door attack on teleprinter

attack, but there are reasons to believe that often this is not the case. 'Bugs', 'taps' and other methods can be used: in the days when electromagnetic teleprinters were used to prepare plaintext messages for input to cryptos, attackers would sometimes try to monitor the current drawn by the punch magnets (Figure 7.10. Knowledge of this would tell them how many magnets had been operated and thus give them a reasonable guess at the character that was being printed. Sometimes they could guess at the complete message, this way. Alternatively it would give them clues with which to mount a known plaintext attack.

Similar attacks have been suggested against the custom circuits used for modern encryption schemes: for instance, it is alleged that the time taken for a particular model of encryption circuit to carry out the enciphering is dependent on the plaintext input. If we can detect these time variations, we can proceed in exactly the same way as with the old-fashioned teleprinters. Again, security is only as strong as the weakest link and, in many cases, this is not a cryptographic weakness.

PRACTICAL CRYPTO-SYSTEMS

So far, we have described the algorithms that form the basis for cryptography, but what about their implementation? One important feature of any realisable system is its speed of operation. Suppose we wish to encrypt a video stream in real time. There is no point in selecting an algorithm that cannot be implemented at megabit per second speeds for an affordable price. This is currently the case with RSA, for example at least when run on a general-purpose personal

Table 7.2 Combination of DES and RSA: Alice wishes to send a high-speed encrypted message to Bob, without setting up a key management regime

Bob	Alice
Bob publishes his public key	
	Alice selects a DES key to encrypt her message
	Alice encrypts her DES key using Bob's public key and sends this and her DES-encrypted message to Bob
Bob uses his private key to decrypt Alice's DES key and then uses the latter to decrypt the message	

computer, where speeds are limited to a few tens of kilobits per second. Special-purpose hardware can raise this to nearly half a megabit per second. The ever-increasing power of computers will cut this time down, for any given length of key, but we must remember that key lengths must also increase in order to keep ahead of the increasing speed of attackers' machines.

Private-key systems such as DES are much quicker, 100 times in software and at least a further order of magnitude in hardware, but, of course, they have the problem that we need to exchange keys via a secure method.

The solution is to use public-key methods at the beginning of a transmission to send encrypted keys from private-key systems such as DES, decrypt these keys and then send the rest of the transmissions using DES (Table 7.2).

AUTHENTICATING A MESSAGE

Security is not just about making sure that messages are not received by people who should not have access to them. A moment's reflection on the practical transactions in our daily lives will remind us that often, particularly in financial operations, we are just as concerned about knowing who is taking part as we are about protecting privacy. Frequently we require to know the 'provenance' of a part of a transaction: where did it come from, is it still valid, is it from whom it says it is? This 'authentication' of a transaction requires a slightly different approach from security, but one which can still make use of the techniques mentioned above.

Authentication can be looked upon to consist of two parts:

- Proof of identity: a reliable witness who can testify that you are who you say you are, or some unique attribute such as a fingerprint or a signature
- Protection against forgery: a means of stopping anyone copying your signature or added to or altering any message you have sent.

The techniques involved therefore require us to find a reliable identification procedure and also some way of making sure this is not forged. Let us first look at forgery.

DIGITAL SIGNATURES

Consider a simple example: Alice wants to buy 1000 shares in a company called Bobco, and wants to do so via an investment trust called Charlietrust. This is a valuable investment and simple prudence dictates that each of the parties needs to protect itself from malpractice within any of the other organisations. Suppose Alice wants to make sure that Bobco receives her message untampered with. Figure 7.11 shows how.

First she generates her message and then passes it to a program which summarises it in a particular way, that we shall discuss later. She then encrypts this summary with her private key and appends the encrypted result to the message as a 'digital signature'.

When Bobco receive it, they reverse the process, using an identical summarising algorithm as used by Alice, together with Alice's public

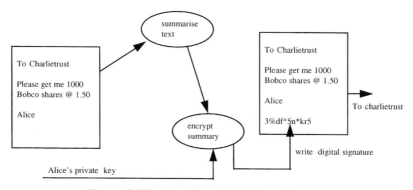

Figure 7.11 Applying digital signatures

TECHNOLOGIES FOR A WICKED WORLD

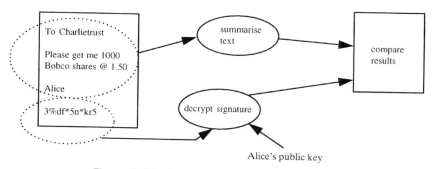

Figure 7.12 Reversing the digital signature

key (Figure 7.12), thus regenerating the summary. This would not be possible if Alice's private key had not been used to encrypt the summary. There are a large number of ways in which we could summarise the contents of a document: count the number of words, assign each letter a numerical value and compute the sum of each line, and so on, but if we are to produce a summary which is resistant to forgery, then we have to use a subset of the possibilities, one which has specific properties. Such a subset are the so-called 'cryptographic hash functions'. These include functions that operate on the data in a way similar to the Feistel functions, of which DES is an example, splitting the data into halves and successively shuffling the halves according to subkeys, but unlike DES they result in shorter, usually fixed-length outputs (greater than 128 bits, typically). The hash functions have properties that make them useful for digital signatures:

- Although there are a number of possible inputs corresponding to a given output, it must be difficult (i.e. expensive and time consuming) to work back from an output to an input.
- Given a specific input and its corresponding hash output, it must still be difficult to find another input that produces the same hash output.

DIGITAL CERTIFICATES

Another aspect of security is that of 'rights' or 'privileges': is the password still valid, should someone still have access to a database or to financial credit? It is all too easy for someone to claim rights without being entitled to them, and this is not simply a technical issue (see frame).

> ## THE DIAMOND SCAM
>
> Jewellery-quality diamonds are highly vulnerable to theft before they are very valuable even in small quantities. Thus, they are kept under close guard and released only to couriers who can repeat the current secret passwords arranged between the buyer and seller. A few years ago, a New York diamond house received an anxious phone call from a purchaser to say that the codes had been compromised, possibly for fraudulent intent, and as a matter of urgency a set of new codes, which he repeated over the phone, were to be used for that day only. Shortly afterwards, a courier arrived, repeated the new codes, was given the diamonds and disappeared, never to be seen again. Of course, it was the crooks who phoned through the codes.

There are two distinct approaches to this problem, which are based not so much on technical issues but more on human attitudes to whom one should trust. In one, which might be described as a 'trusting citizen approach', we make use of a 'trusted third party' or 'key distribution centre', which possesses some status as a reliable and secure organisation which will not cheat you or imperil your security. Suppose, for example, Alice wishes to communicate with Bob. She chooses a key to use for the encryption between them and, using another key that she shares with the key distribution centre, encrypts the first key, together with the destination address ('Bob'). The Distribution Centre decodes the message and then recodes it using a key it shares with Bob. This is the bare bones of the technique and, as it stands, is vulnerable to what is known as a 'replay attack' where a third party can piggyback her messages onto the legitimate ones, but the basic principle is sound. However, it does rely on a belief in the integrity of the Distribution Centre.

In an alternative approach, which we could call 'justifiable paranoia' ('Just because you are paranoid, it doesn't mean they aren't out to get you'), there is no central Key Distribution Centre. This is comforting to people who do not trust governments, banks or other quasi-official bodies. Instead, users maintain a 'key-ring' of several public and private key pairs, rather than just one pair, in case they believe some to have been compromised. They also hold a key-ring of public keys of their correspondents. They keep a record of how good they believe these keys to be, in the sense of how much they can trust them to belong to the stated owner and whether they are likely to be compromised. The most trusted public key would be a newly generated one personally handed to you by the owner. Slightly less trustworthy would be one passed on to you by a mutual friend. This method allows for the generation of networks of trust, with trust being represented as a continuous value from 0 to 100%, rather than as a simple binary value of 'trusted' or 'untrusted'. This is probably closer to real life than the alternative. However, just as with life, it is doubtful as to whether it would scale across large communities.

The same principles (and the same divergence of approach) are adopted in the control of keys themselves. In the 'official' method, the identification of a public key with an individual and within an expiry date, together with a unique identification number and a digital signature, is generated for the owner by a 'certification authority'; this body will be registered with the Internet Policy Registration Authority as providing a specified policy on registration (foe example, a company could register as certifying certificates for its employees). The current validity of the certificate must be checked every time a message is sent. The whole process is hierarchical and procedural and, therefore, does not find favour with the libertarian/paranoid lobby.

PORTABLE SECURITY – THE SMARTCARD

Suppose we have a security system, one based on key encryption and possibly some biometrics. How will we use it in practice? Rule one must be that we should try to avoid having long, difficult or numerous passwords to remember, because we shall either forget them or write them down. Another rule will be that we should not use our biometric 'in-clear' over the network, for fear that it be intercepted and illicitly used by a third party. We want to pass over the network an unforgeable signature that can be used to confirm that it is us and no-one else who is there, during the current transaction. One way to do this is to carry a personal identity token in the form of a 'smartcard'. As shown in Figure 7.13, smartcards are computers encapsulated within credit-card sized plastic bodies. The processors are designed to run rather slowly but thereby consuming little electrical power even when running at full speed, and the entire card consumes pico amps (for memory retention) when not in use. The card has fixed, read-only memory which contains programming instructions and data such as the user's private key, although the latter may instead be embedded

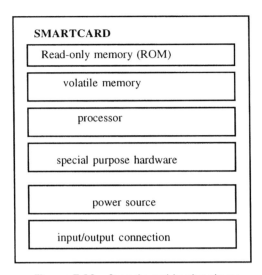

Figure 7.13 Smartcard technology

in any on-card, special-purpose hardware, such as an encryption chip. A number of input/output options are possible: current cards usually make contact with other devices by means of electrical connectors on the edge or surface of the card, but magnetic, capacitive or radio connection can also be used, as can optical methods.

Smartcards can store generally useful information; for instance, they can hold a list of telephone numbers that we can invoke by a short code dialling code (1 = 'mum', 2 = 'work', etc.). In the case of GSM mobile telephones, they can be a full personalisation module that holds our own identity code: without one, no GSM phone will operate; with it, the mobile can send out your identity to the mobile network, giving access rights and allowing charges to be attributed to your bill. Cards can also hold the crypto algorithms (for example, a public key algorithm) and can contain your personal private key, so that you never need to remember it. You can use one to personalise, and provide a secure channel from any telephone or computer terminal that you have access to, without needing to key in a password. The key is 'burnt into' read-only memory, so that it cannot be changed maliciously, and the memory chip 'trapped' so that any attempt to break into it results in the key being erased. Given this level of personal security, we can even use if for money.

A study carried out in 1995 identified nearly 200 organisations actively involved in developing equipment for smartcards or their

Table 7.3 Smartcard applications

Application area	Personal authentication	Portable data files	Portable personality	Electronic purse
Transport	×	×	×	×
Retail			×	×
Finance	×	×		×
Security	×			
Communications	×		×	×
Home	×	×	×	
Utilities		×		×
Health	×	×		
Roads	×	×		×
Licence	×	×		
Education	×	×		
Passports	×	×		
Social Security	×			×
Employment	×	×		×

Source: M. Meyerstein, BT Labs

integration. There is vigorous activity in the area of international standards for card architectures and basic parameters such as working voltage and specification of the electrical connection, as it is seen that a major cost-reduction opportunity exists if the number of variants can be reduced. The study identified a very wide area of application, as shown in Table 7.3.

PROOF OF IDENTITY – BIOMETRICS

One vitally important aspect of security in a networked environment is proof of identity. This was not a problem when people dealt face-to-face, but how do we identify someone at a distance? We have seen with digital certificates that public/private keys can be validated by a third party, but there is still no way that these can be uniquely associated with an individual, without something extra.

What we need is something that identifies a person through a unique biological characteristic. These 'biometric' measurements rely on the identification of some pattern of habitual behaviour or physical characteristic that is more persistent in some way to an individual than to people in general. Not only must the measurement be able to identify individuals, it must also be possible to prove that this is so. This tends to favour physiological parameters, such as fingerprints or retina images, which are precise and relatively unchanging throughout time or situation, rather than 'habitual measures, like voice of handwriting, which are learned habits rather than genetically 'burned in'. Nevertheless in practice, the measurement must be cost-effective and its administration acceptable to the user, and this sometimes means that habitual features have an important place.

Identification by biometric means is a particular example of the general class of 'pattern-matching problems'. Given an example from an unknown subject, there are a number of techniques whereby this can be compared with existing patterns from a number of subjects, in order to find out which subjects' patterns it is most like, and hence provide a best guess at the identity of the individual. There are numerous techniques and these have been applied to a variety of biometric types using a range of measurement technologies.

Basic Principles – Linear Classification

One such technique is 'linear classification'. Since most of its principles can be applied to other, more complex techniques, we shall look at it in a little detail. Suppose we decide to use a simple handwriting recognition scheme. Consider the problem of recognising the authorship of the handwritten letter 'W' in Figure 7.14.

We have decided to measure two 'features' of the letter: the average slope of the downstroke AB, and the speed of the upstroke CD. Suppose we have several examples of W's written by each of N people. They will not all have the identical value, neither across all writers nor even for each individual. Suppose we plot the values on a graph as shown in Figure 7.15.

Each different shaded shape represents a ring around most of the

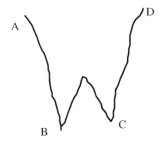

Figure 7.14 A handwritten letter 'W'

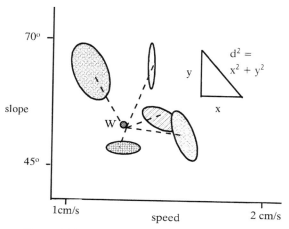

Figure 7.15 Distribution of individual variations

values for any one writer. We see that, taken across all writers, the value of the slope ranges from about 45° to 70°, and that of the speed between about 1 cm/s and 1.5 cm/s. We also see there is a mystery writer, 'W'. To make the figure clear, we have deliberately chosen a value of the two features for W that is far from the centre of the values for any one writer, but W must be one of them. Who is 'W'?

Our first thought is simple to measure the lengths of the line between W and the centres of each writer's spread of values and choose the writer who is nearest. On a two-dimensional graph, this distance equals the square root of (x distance squared + y distance squared), according to Pythagoras's theorem. But this would not do. Suppose we simply changed the horizontal sale from cm/s to m/s, or the vertical from degrees to radians? The relative distances would change and we might find that we now identified W with someone else.

Rather than use arbitrary scales, we have to select scales that take into account the relationship between the spread of values for each feature across all writers, compared with the average spread for each feature within any one writer. We define a ratio, conventionally called the 'F ratio', where

$$F = \frac{\text{range of values of the measured feature across all features}}{\text{average range of individuals}}$$

The F ratio is used to weight the calculation of the distance between the unknown example and the various possible categories (in the handwriting example, the different writers). A further practical significance of the F ratio is that it is usually easier to establish confidence in a component of a system if its F ratio is high, as we shall see in the following examples.

HANDWRITING RECOGNITION

Authorisation by means of a signature is almost as old as writing itself. We have previously briefly mentioned measurements of direction and speed and there are a large number of patents that employ variations of these. Some operate only on the static signature (i.e. direction), while the newer approaches increasingly emphasise the dynamic approach. Handwriting possesses some of the problems associated with other 'habitual' measures: the F ratio is not particularly favourable; we need to separate out the variability that each individual has in his own signature from the features that separate individuals. There is also a slight technology problem: to capture the signature, we need either a relatively expensive tablet to write on, or a generally cheaper, wired pen which, however, is more liable to be broken (or stolen!). One reason why the tablet is expensive is that it may need to be sensitive to 'air strokes', the movement of the pen when it is *not* in contact, as this is also a distinguishing feature of signatures.

Fingerprints

This is a biometric measurement that antedates electronic technology. There is little doubt that fingerprints offer an extremely effective way of identifying individuals uniquely, provided the print can be captured reasonably completely. The F ratio for fingerprints is almost infinite (individual changes are approximately zero, unless fingers are deliberately or accidentally profoundly damaged). It is not necessary to have this property, but it does make it easier to justify the security of the biometric.

In principle, an image of a fingerprint could be captured by a TV camera, transmitted over a communications link and checked against a database. In practice, however, there have been a number of problems with automated, remote access systems: because of the high density of lines within a print, a large amount of data is required, and a not insignificant amount of processing power is needed to align the data if the print taken is only partly captured or poorly outlined. Fingers must be pressed against glass screens in order to position them correctly for scanning; the screens can easily become dirty.

Recognition of Hands

A very simple, but quite reliable, biometric is the three-dimensional shape of the hand. This can be measured quickly and accurately, using basic technology (Figure 7.16).

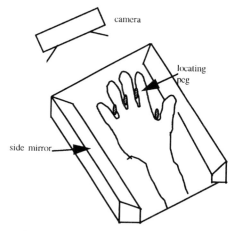

Figure 7.16 Recognition of hands

A system like this is already proving its worth with the US Immigration Department, for immigration control to the US. People already legitimately in the USA can acquire smartcards that hold an encrypted version of their hand measurements. If they leave the country and subsequently return, they can be automatically monitored against their card to see if they are who they claim to be. One of the major advantages of this technique is that only a small amount of data is required to specify the biometric – one system quotes nine bytes.

Voiceprints

The name 'voiceprints' has been given to the technique of 'speaker verification', because it has been claimed that each individual's voice has unique characteristics that are analogous to the uniqueness of fingerprints. This claim is highly misleading and has never been

Table 7.4 Comparison between fingerprints of 'voiceprints'

Fingerprints	'Voiceprints'
Genetically determined	Genetic determination has not been proven. There is a degree of correlation between voice and physiology (e.g. length of vocal tract) but the former is not rigidly determined by the latter. Voice may be learned; certainly, accent is acquired, not inbred
Cannot be altered by will; do not change with age	Voice changes with situation; voice can change dramatically with age, state of health, location
Individual's prints are identifiable, discrete lines that can be seen to retain the identical pattern under all circumstances	Individual's voiceprints comprise complex 'features' that are flexible, vary from utterance to utterance, and have values that are sometimes shared with others

convincingly demonstrated. There are wide differences between the two (Table 7.4).

The real problem that the advocates of speaker verification have to deal with is the fact that the F ratio for speech does not appear to be anything like as great as that for fingerprints, or, at the very least, is extremely difficult to measure, for any of the features that go to make up the various systems on offer. Thus, it will be difficult to estimate the reliability of speaker verification schemes, and so it is unlikely that they will be trusted on their own for critical applications. Nevertheless, in communications, speech is one of the easiest data sets to collect and there are indications that speaker verification does give a modest degree of protection: we could say, for example, that from a population of 1000 voices we could reasonably claim that we could narrow the possibility down to 50 or even less, with a 'good chance' of being correct. If, for example, the authentication procedure comprised asking someone to speak their personal identity number (PIN), voice recognition would thereby add significantly to the security of the process over that of simply keying in the PIN.

Regarding the details of voice recognition, there are a large number of parameters that have been tried: voice pitch, frequency spectrum, vocal tract resonances ('formants') and derivations thereof. Some are more resistant than others to the noise and distortion problems of communication networks, or require less computational activity.

Keyboard Idiosyncrasies

Another example of a statistical parameter that is a 'natural' candidate for use over the network is keyboard statistics. Increasingly, at least for the next few years, keyboards will be the predominant way of personally passing information across the network. It would seem logical to investigate whether we can be distinguished, one from another, by the rhythm with which we use the keyboard.

There is a respectable history behind this: during the Second World War, British warships would set off from port after landing the Morse code operators they had used when they were in harbour. These operators would continue to transmit messages from shore, knowing that the German interception units would recognise their 'fist' and therefore believe the ship was still at anchor.

Application of this to automatic recognition of keyboard style has been achieved in a number of ways: one method measures the press and release times of the keys when users typed names. From this was calculated the time between successive presses and also the time the keys were held down and these were used as classification parameters. The system was trained using a neural network (see Chapter 6) although, in principle, other classification strategies can be used.

In assessing performance of classification schemes, we have to look at two types of errors, 'false recognition' errors, where the system wrongly identifies the unknown user with a specific known user, and 'false rejection' errors, where it fails to make a match when it should do (Figure 7.17).

Most schemes allow us to set recognition thresholds that determine where on a graph of false recognition/false rejection we wish to operate.

Usually, if the access is to critical procedures, we will choose to operate with a low false recognition rate and thus sometimes will expect to reject, wrongly, access attempts by legitimate parties (although we cannot have too high a false rejection rate, otherwise the system will be unusable).

In the case of the handwriting example discussed above, results have been reported on a test population of 50 and the threshold was set so that no false acceptances were made. This resulted in a rejection rate of 50%, i.e. on average, the users had to key in their names twice. This is probably acceptable.

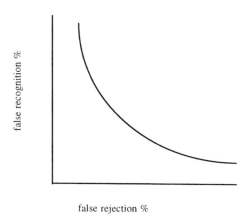

Figure 7.17 False recognition and false rejection errors

Retina Scanning

Recently, significant success has been reported in practical measurement of the pattern of veins on the surface of the retina and the use of these as a unique 'retina print' for individuals. Indeed, retina scanning is one of the front-runners in the quest for a reliable, non-invasive biometric. The user simply looks at a TV camera and an image is captured and processed on a medium-range computer workstation costing a few thousand pounds. The technique has the obvious advantages of fingerprints – a very high F ratio – as well as being non-contact and providing a better-quality image. One system quotes a stored pattern of 256 bytes and a 'cross-over error rate' (that is, when the false acceptances equal the false rejections) of better than 1 in 130 000.

Biometrics and Remote Access

One obvious use for biometrics is in allowing access to restricted areas, and some of the earliest applications of the techniques described above were just for that, the provision of 'keys' to locked doors. The situation is slightly different when we consider using biometrics for access to resources at the other end of a network. In a standard 'doorway access system', you have to present the feature to be measured (e.g. your hand) to the system directly; in the networked scenario, the distant 'doorway' is presented with a digital signal that purports to be a measurement of your biometric. It is easier to forge a digital signal than it is to forge a human hand! For instance, what is to stop someone intercepting one of your access attempts, capturing the data corresponding to your biometric measurement, and reproducing it at a later date?

Sometimes, this risk is worth taking, if the value of the transaction is low and convenience is more important. In this case, simple zero-cost entry methods such as voice recognition are acceptable. If the transaction is critical, very sensitive (e.g. in the case of medical or other personal records) or potentially lucrative to criminals, then something stronger is required. The usual solution is to encrypt the biometric data, using the techniques we discussed earlier. In many cases, it will be convenient to use a smartcard to hold the encryption details, and it may also be useful to hold, in encrypted or trapped form, the biometric details, in case the card forms part of a local access scheme such as the immigration control example given above.

ELECTRONIC COMMERCE AND ELECTRONIC MONEY

Perhaps the most important use for networked security is not the preservation of state secrets but rather the provision of a safe way to protect transactions involving money. Already there are large-scale, secure networks of handling interbank transfers. These resemble military networks, with high volumes of data, operating between secure premises and under close supervision. But there is another place where transactions involving money are taking place – the electronic marketplace consisting of telephone and Internet purchasing, a 'retail' environment which involves lots of (frequently very small) transactions carried out at a distance over relatively insecure networks between parties who may never have dealt with each other before and may never deal again.

Figure 7.18 shows one possible arrangement for implementing such transactions. The purchaser and the 'shop' together identify the goods to be purchased, their cost and the identification code. The purchaser then uses an arrangement from a bank or other certification authority to generate a validated and encrypted message that she attaches to the order form for the goods. She sends this to the shop, which can read the order and feed it into its order processing system. It cannot read (or subsequently forge) her validation details. It does, however, attach its own digital verification to the order and sends the combined message to its bank. Its bank then sets up a transaction between itself

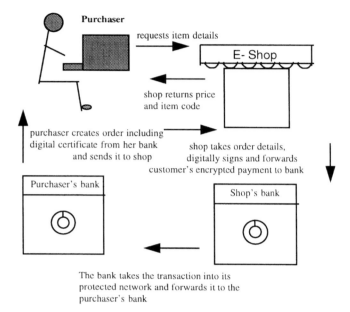

Figure 7.18 Electronic commerce payments

and the customer bank, wherein the latter can decrypt the customer's certificate and the shop's bank can receive the payment due.

ELECTRONIC CASH

The above process is an effective way for trading at a distance for relatively 'big ticket' items, that is items costing several pounds, but it is not ideal for very small transactions, for the same reason that credit cards are not used for them today – every transaction incurs processing costs and for every transaction there will be a consequent charge. The alternative is cash, which incurs no overhead, and it is a fact that nearly 50% of physical transactions fall below £5 and consequently are paid for this way.

To provide an electronic alternative, we need to have some way of downloading, in bulk, a sum of electronic money, thus incurring only a single handling cost. This is yet another application for a smartcard. First, we must maintain a supply of cash in our card or 'electronic purse'. Figure 7.19 shows how this might be achieved using an advanced telephone.

The cash is now held on the card, which together with a convenient

TECHNOLOGIES FOR A WICKED WORLD

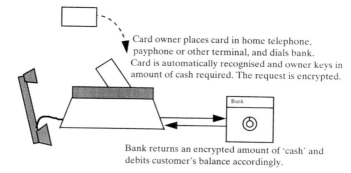

Figure 7.19 Acquiring electronic cash

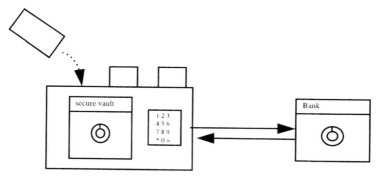

Figure 7.20 Paying with electronic cash

'reader' acts as an electronic wallet. The reader is the size of a pocket calculator and can hold balance information, etc. It is also used to 'lock' and 'unlock' the card, using a personal PIN, to allow nominated sums of money to be taken off the card, for example in a shop. As shown in Figure 7.20, shops have electronic terminals into which the cards can be inserted and which can remove the amount of money appropriate to the cost of the goods purchased. The 'cash' removed is electronically encrypted and held in a secure 'vault', for low-cost bulk downloading to the shop's bank, where it will be credited with the corresponding amount.

CONCLUSIONS

In this chapter, we have looked at a number of issues regarding security and the possibilities for electronic trading and general information protection. In terms of time-frame, it appears that the technology has now come of age for providing secure solutions that can be administered without requiring the cumbersome infrastructure that once bedevilled secure transactions. It seems as if many of the major problems have been solved – but we must always realise that technology is not the sole province of the good; one lesson we learn from traditional security that will never be obsolete is that each measure has its countermeasure and continuous development will be required. Furthermore, successful crime seldom results from superior technology in the hands of the criminal; usually it happens because of human failings. Any technology that is a genuine contribution to the protection of property or information must be designed to make misoperation as unlikely as possible; it should be simple to use, difficult to misuse and something that its users can trust and respect.

8

The Stakeholders

SERVICES AND STAKEHOLDERS

Naturally, the way that one looks at a set of products or services is heavily influenced by the position you start from, in particular whether you are a supplier or a customer. This is just as true in communications and computing as it is elsewhere, perhaps even more so, because of their high technical content. These distinct viewpoints can, at worst, create conflicts of interest, and at best can lead to confusion. In the domain of computing and telecommunications, one area where this is particularly so is that of 'service'.

As shown in Figure 8.1, the supplier tends to think of the technologies involved, on top of which are constructed a limited number of 'service elements' that can be combined to support any number of applications. Suppliers use the service element concept in order to allow them to decompose an application into the components needed to support it, without having to predicate this on any specific set of technologies. Services, or service elements, are abstract technologies.

This is useful to service providers, but it does not reflect the way customers approach the concept of a service. They view it first of all from the position of its benefits to them: what it does (sometimes including how it does it – its 'technologies'), what are the payment terms and what support infrastructure exists to help them get started and then continue to use it. In this chapter, we shall look at 'service' from the latter viewpoint – that of the user – although we shall

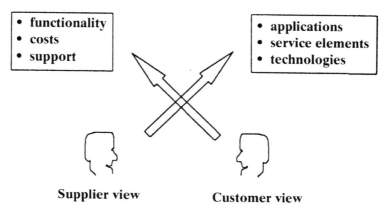

Figure 8.1 Two views of 'service'

approach it, when appropriate, in terms of the technologies that might support it. Because we are looking at aspects that are driven by markets, legalities, regulation, etc., it is even more difficult to predict what is going to occur than with technology alone. Moreover, whereas technology can usually act as an enabler to new services, most of the issues we shall touch on are barriers, whose removal depends on a combination of many factors. This chapter is therefore biased more towards issues than to solutions.

The direct 'customer', the person who pays for the service, is not the only stakeholder in it. Sometimes the customer may not be the user – public services are often procured by government or local authorities, on behalf of individuals. Sometimes there are issues of unfair competition that impact on other would-be providers. No enterprise as large and as pervasive as the telecommunications or computing industries can expect to be immune to 'interference' from influences beyond their user-communities. The stakeholders include:

- Regulators, who intervene in free trade to promote a government objective (usually 'competition')
- Creators of content, who have a right of ownership and a duty not to defame
- Citizens who have social rights – to privacy, decency, employment, etc.
- Citizens with right of protection against crime or offences against the state
- People with special circumstances, e.g. with disabilities.

MEMO

TO: REGISTRY

FROM: ANDERSONIAN LIBRARY

USER: Azhar Malik CARD DOES NOT WORK PLEASE ISSUE WITH A NEW ONE.

THANKYOU

WENDY BEAR

REGULATION

Increasingly in every part of the world where there exists an up-to-date telecommunications infrastructure, it has been, or is in the process of becoming 'deregulated'. That is, the market for telecoms services is no longer the monopolistic privilege of a single operator who in the past was usually the state. The justification for this new approach is a belief that better, cheaper and more innovative services will be provided where an atmosphere of competition prevails. (It should be noted that not all analysts agree totally with this assessment; most believe that cost reductions are likely to be achieved but there are a number who deny that quality of service or innovation are thereby necessarily improved.)

The problem of ensuring fair competition is not a simple one and the approach adopted by governments to achieve this is highly interventionist. The approach most commonly adopted is to introduce an independent authority, a 'Regulator' (FCC in the USA, OFTEL in the UK, for example), whose job it is to oversee legislation and monitoring of the evolution of telecommunications to the new model. Naturally, the competition issue is the predominant one that faces the regulator.

The task of regulator is not a job to be sought by anyone who wishes to be universally popular, nor is it a job for someone with too few or too many opinions. It is a job fundamentally built on contradictions – how can one create an 'interventionist free-market'? In theory, at least, regulators should be working towards making themselves redundant. It has been said by a number of leaders of telecommunications companies that this is not the way they will work. Some have also said that the pace of technological change is so great that it is fundamentally impossible for regulation to keep up, and it will be repeatedly guilty of making nonsensical statements.

Nevertheless, regulation is a part of the apparatus of the freeing of the market, and telecoms companies, whatever their views, must adapt their business models to acknowledge this. Not only do they need to create different commercial architectures; there is also a need to consider the technical issues that are required to support the new commercial environment. In the sections that follow we shall look at some examples.

Fair Access

Because of the monopolistic legacy of telecommunications in most countries, it is difficult for a new entrant to commence operations on equal terms.

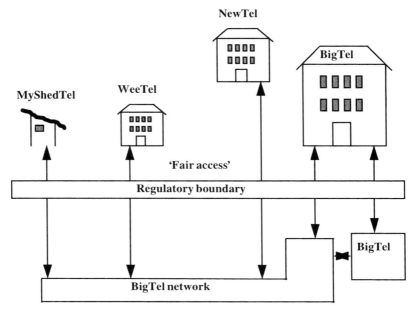

Figure 8.2 Evolution of competition

For instance, in Figure 8.2, we imagine 'OldTel', the state-owned monopoly, begins trading after privatisation under the more attractive name of 'BigTel'. The name is an accurate reflection of its market share (initially 100%), its central network and its local connections to customers. Enter 'NewTel', a new competitor. How does the regulator ensure that NewTel, which has no customers, no historical infrastructure financed in the past by the state, etc., can compete fairly? There appears to be no one universally agreed solution. In the USA, for example, the competitive element was created by separation of the state monopoly into a long-lines and a local-loop pair of markets. Companies could not own both the trunk delivery and the delivery to the customers from their premises to the local exchange. This makes some sense in the USA, where effective competition in the local loop could be made to happen by permitting the fairly extensive cable company networks to carry telephony. Then, the competition on the long-haul circuits (which is not so dependent on historical investments) would also be opened up, as a separate business.

In the UK, on the other hand, with little local loop competition readily available, because of the low level of cable activity, it was deemed preferable to leave British Telecom intact but to require the company to carry traffic for other companies, across its local and trunk network, on the basis of fair access. As cable networks grew and were permitted to offer telephony, British Telecom was also required to carry the trunk traffic they generated.

Despite the rather different ways of working, there is, however, implicit in both cases the concept of a 'regulatory boundary' which separates the interior of a telecoms company's network from its exterior. To understand this properly, it is important to realise that this boundary is indeed 'regulatory', not identifiable with any particular technical architecture, although the technical architecture will, in some important areas, be affected significantly. In Figure 8.2, BigTel is the only company that has an extensive network. Over this network it runs a number of services that it sells to its customers. Now onto the scene come 'NewTel', a sizeable competitor but without much of a network in this territory, 'WeeTel', a smaller innovative company that may wish to sell new added-value services, and 'MyShedTel', whose name implies that it is a very small, perhaps community-based company, which offers nothing especially new, nothing outside its own small region, but still the holder of a licence from the regulator that gives it full rights to be a 'public telephone operator' (PTO). All of these companies (including of course BigTel) require access to parts of BigTel's network. It would seem reasonable to provide this access on 'fair terms'. The regulatory problem is in how to define 'fair'.

First the regulator must determine which part of BigTel's network is common to all of them. This is not as easy as it sounds. In the UK case, for example, BT and some other telecoms companies deliver to end-customers over BT's local cables, whereas cable companies use their own networks and only connect to BT (or another long-distance carrier) at the exchange, as do cellular radio providers. Once the regulatory boundary has been drawn, then the regulator will need to ensure that BigTel's operations in running the network 'below the boundary' (see Figure 8.2) will be sufficiently separated in terms of costs and profits to prohibit unfair cross-subsidy. BigTel will also have to make sure that their operations and maintenance systems can preserve this distinction. For instance, network management problems and the cost of their resolution should be clearly separable into those caused by the BigTel network and those due to problems at the interconnection across the boundary. As services get more complex, this will also apply to help desk functions as well – if a failure occurs somewhere in the supply of an interactive TV service, how are the restoration costs to be divided up between the various parties?

But the regulator's task is not over once the regulatory boundary is defined. There is still a need to define the terms and conditions for completing the connection across the boundary. Does 'fair' mean 'at cost' at 'equal price' or what? In the US case, there was some debate about whether the connections between the AT+T trunk exchanges and the local exchanges should be cheaper than the connections to the new long-haul competitors, since the former, for historical reasons, often shared a building with the local exchange whereas the others were often kilometres away. Alternatively, can one offer bulk discounts, so that BigTel and NewTel can effectively freeze out WeeTel and MyShedTel? There is no easy answer and the regulator has to steer a pragmatic course between the competing interests involved, often to the annoyance of the companies that lose out.

There is one aspect of the regulatory boundary concept that can have a particularly acute impact on technical architecture: consider the case of a radically new service. Let us imagine that engineers at BigTel have come up with a solution to the problem of recognising people's faces, and they want to use this as a way of providing a 'user authentication service', for example to provide secure electronic banking. As shown in Figure 8.3, the service requires public terminals that include a TV camera, and a system associated with the network that carries out the recognition and any administration involved. BigTel have two choices: they can choose to put the service above, or below, the regulatory boundary. What are the implications? If they put the service below the boundary then, if the regulator spots this, the service must be made available to New Tel, WeeTel and MyShedTel on a fair-access basis, losing at least some of BigTel's exclusivity.

'Right', says BigTel, 'I'm not going to put it below the line; I shall sell it as a "value-added service" above the line, as part of my above-the-line company. I don't need to give anyone else access to it, so there!'

'OK', says the regulator, 'but I shall expect you to show clearly that the cost of the service is not being partly funded out of your below-the-line activities. Oh, and by the way, when you calculate the cost of the service, you will include the cost of providing access into and out of your regulated network, too.'

The last remark can have serious implications. The regulator is referring to the dotted lines on the figure, going into and out of the regulated network. Suppose, instead, it was NewTel that came up with the idea of the authentication service: if a user accesses the

THE STAKEHOLDERS

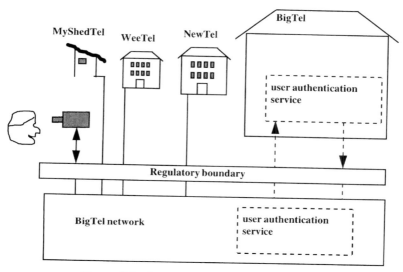

Figure 8.3 Position of added-value services

service, it will, by necessity, because BigTel owns the local loop, come via the BigTel network and BigTel will charge NewTel for providing the access and any further actions (such as the NewTel service dialling out to a bank). So, when NewTel calculate the cost of their authentication service, they must include any fees paid to BigTel. Now reverting to our original supposition that BigTel introduces the service, what the regulator is demanding from BigTel is that is charge itself the equivalent fair access charge, across the boundary. What this means in practice is that BigTel services will probably be priced higher if they are above the regulatory boundary than if they were below it.

(Just to make this clear, total costs of networks and services do not go up. If the service is implemented above the boundary, it is a cost to the value-added service provider and a source of revenue to the network, therefore the network charges are lower – but this saving is spread over all the customers of the network, including those who get service from NewTel, WeeTel and MyShedTel.)

Clearly, this has a major effect on technological strategy. This becomes particularly acute when sizeable rival networks have eventually developed. A sensible strategy for WiseTel to pursue (see Figure 8.4) is to introduce its new service platforms outside the regulated network, accepting that its value-added services company will have to pay connection charges to that network but thereby keeping exclusivity

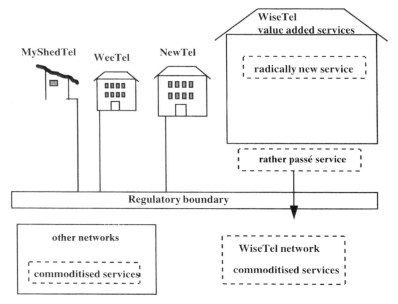

Figure 8.4 Service migration

and gaining revenue. On the other hand, WiseTel's network company also wishes to make money and it will only do so if the network it offers is sufficiently attractive to other customers, including acting as a bearer network for other telecoms operators. At some stage in its life-cycle, what was once a radical new service becomes a commoditised, basic feature that all competitive networks will be expected to have. WiseTel's policy must therefore be to determine when it should migrate a service across the boundary and thereby make it available to all, on a fair-access basis.

This means that advanced services will begin outside the network core, another example of our earlier observations that peripheral intelligence will grow. By running the new services in a value-added, restricted domain, companies will be encouraged to experiment with non-standard technology, increasingly using the rapidly evolving and less reliable computer systems, rather than homogeneous, extensively tested, ultra-reliable public network equipment.

Meanwhile, the business directors of the telecoms companies will be trying to achieve maximum return, not just over one business, but over the sum of the value-added business and the network business.

Number Ranges and Portability

Telephone numbering schemes have evolved to meet the changing patterns of call rates and routes, but in the past they were designed under the regime of a monopoly supplier and a single type of network. Now, with unprecedented growth in call rates and new premises, the emergence of new competitors and new types of networks, notably mobile ones, regulators must become involved with the debates on numbering plans.

Telephone numbers are not just 'numbers', in the sense of meaningless ciphers; their construction contains a significant amount of information to the system and to users:

- for most fixed network users, it locates them within an exchange area
- for some fixed network users, it gives them access to special overlays, such as 0800 service
- to mobile users, it gives a unique identity to the mobile equipment
- to all of these, it indicates which company is providing them with service
- to all of them, it provides an indication of the charging mode/cost that will be incurred by people who make calls to them
- to all of them, it provides some form of 'landmark' that identifies them and allows people to reach them, irrespective of distance.

In becoming involved in the numbering debate, the regulator finds itself, not for the first time, trying to reconcile incompatible points of view. Take, for instance, the debate regarding personal numbering. The proposal is for providing people with a 'number for life': one that they can be contacted on, irrespective of whether they have remained with the same telecoms company or not, or whether they are on a mobile or fixed network, with the assumption that they should be free to change from one to another to suit their convenience.

It is, however, important to realise that, in a two-ended process such as a telephone call, the needs of the other party must also be considered.

In Figure 8.5, A has made a call to B. A is in one telephone zone and B is in an adjacent one. There happens to be an arrangement whereby calls between adjacent call zones are charged at a reduced 'local rate'. A knows this and is therefore happy to call B. Now consider the position shown in Figure 8.6. Another telecoms company has entered the market, and B switches to them. Unfortunately, their zones

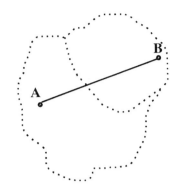

Figure 8.5 Adjacent call-area connection

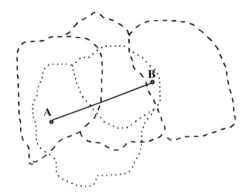

Figure 8.6 Multiple service provider scenario

(dashed lines in Figure 8.6) are different and an extra zone is interposed between A and B. This results in much higher call charges. Because B has not changed numbers, thanks to number portability, A will incur extra charges without necessarily being aware that that is the case.

B does not even need to change suppliers, merely to move from a low cost service, such as a standard fixed line, to higher cost one, for example a mobile one. Through number portability, this move can be concealed, deliberately or accidentally, from the caller, leading the latter to face an unwanted bill.

There are a number of ways round this: there could, for instance, be a recorded message, prefixed to each call, outlining its charging principles. Alternatively, the cost of the call could be split between the caller and the called party, with the called party carrying any excess charge, for mobile access for example.

But although all of this is possible, the regulator must make it happen according to a schedule, because it involves considerable co-operation between competing organisations, and they have to design billing systems and network management to cope with the new complexity.

Homogeneity versus Innovation

In the previous paragraphs we touched on another of the balancing acts the regulator must perform – seeking to obtain consistency of behaviour between competing organisations. This is one of the new difficulties brought about by liberalisation; in times past, telecoms companies had territorial monopolies given to them by the state and their approach to their opposite numbers in other states was by and large highly cooperative. It was in their interests to cooperate closely, because this could only further the growth in international calling, of which they would get a share. Thus, there was an entirely rational motivation to work together to produce standard ways of doing things. This is not the case when regional monopolies are removed and companies are forced to fight for market dominance by differentiating their products from their competitors. What does the regulator do, for instance, in the case of Calling Line Identity (CLI) systems? These systems allow for the caller's exchange to send a short data signal to the called party's telephone, indicating the telephone number of the caller. (It can be withheld at the caller's discretion.) The number is made available before the called party answers the telephone. There are a number of ways this can be done. In the original mode, developed in the USA, the data stream (which comes in the form of modem signals) is injected between bursts of ringing; in the scheme developed by BT, which has to cope with the different ringing cadence of the UK system and which they also wanted to have greater capacity than the American method, the data stream precedes ringing. This means that the BT system and the systems used by cable companies in the UK (which are based on the American system) are incompatible. Rules imposed by the regulator require dominant companies to give 15 months' notice of the introduction of new or modified customer interfaces. This is intended to allow other companies to prepare for its introduction, perhaps even equipping themselves to provide it also. One would be tempted to ask, 'Why 15 months? Is this long enough to respond to change, or too long in that it delays innovation and destroys legitimate market edge?'. Tempted, but for the fact that OFTEL has been asking itself exactly the same question. (See the OFTEL report cited on page 381, 'Selected Further Reading'.)

Regulation and Information, Communication, Entertainment Services

So far in our discussion of regulation, we have concentrated on traditional telephony. However, as we have noted elsewhere, the telecoms, computing and consumer markets are converging onto the so-called ICE services: information, communication and entertainment. (An acronym preferred by the UK regulator is BSM – 'broadband switched mass-market services.) OFTEL have proposed a business model for these activities, which identifies four market segments, which we might consider to comprise a value-chain from source to customer (Figure 8.7).

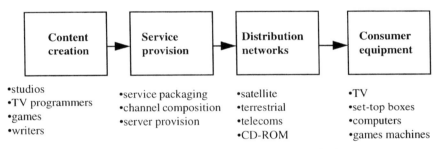

Figure 8.7 OFTEL mass-market value chain

The UK regulator also has views about the state, particularly the competitive environment, in each of these markets:

- *Content creation* is seen to be quite healthy, although there are copyright issues and other barriers to equal negotiation for access rights.
- *Service creation*, which is taken to include the setting up of multimedia data servers, the creation of presentation, navigation, access control and billing for this data, is seen as embryonic but perhaps the key to innovation in BSM.
- *Distribution networks* are telecom, cable, satellite, terrestrial systems, also including CD-ROM, videotape and other physical means of publishing. Here, the regulator has some cause for concern over market dominance by a few players.

- *Consumer equipment* includes TV, computers, satellite dishes and set-top boxes, games machines, etc. There is a lot of competition, but some examples of market dominance.

The regulator has formed considered views about the issues surrounding these four categories. Whether these views are correct is less important, in the short term at least, than the fact that they are held, because they are likely to become policy. One of the problems that regulators have, of course, and one which is acknowledged by OFTEL, is that regulation is not coherent across them. Although the technology has converged across the computing, communications and consumer domains, the rules have not. For instance, a TV set-top box, to receive satellite signals, does not need regulatory approval, whilst one intended for a video-on-demand service, and thus connected to the telephone network, does, in theory at least, require to undergo an expensive and protracted approvals procedure, although the technology used is virtually identical.

The regulators are concerned about the dominance of any supplier at any point in the supply chain and also that dominance in one part might be used to force customers to choose the supplier's service in other parts of the chain. For instance, a company could buy up particular sources of TV programmes and deliver them only over its own network, or via its own consumer equipment, or insist on any other party buying a bundle of programmes rather than being free to select those it wanted. There is also, for instance, a particular issue, again with set-top boxes, in the selection of the decryption components. It is unlikely that users will wish to deal with a multiplicity of standards that involve them purchasing more than one box (or even an expensive plug-in). Therefore, possession of the encryption technology will create a market dominance that could be open to abuse.

Charging for Content

Regulators have also been known to take an interest in the methods employed by distributors (the telecoms companies) to levy charges for the content as well as the delivery. At present, the only method of so doing is via the 'premium rate' services which are prefixed by a special group of numbers (e.g. 0891) and which thereby incur a higher rate of charging for the call. This excess charging is split between the telecoms company and the service/information provider who is responsible for the creation of the content. This must be considered an expedient: there can be no obvious connection between call duration and the value of the information on offer. Regulators are principally concerned with protecting against concealed, excessive charging, whether legally incurred or not (see frame), but they are also eager to be seen taking a responsible attitude to barring access to unsuitable services. One way of achieving the latter is by requiring the telecoms companies to provide a PIN service to which access can be made available only if the would-be user also keys in a secret password.

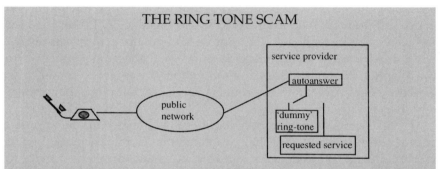

THE RING TONE SCAM

This is a relatively easy swindle to set up; it has been deployed at least once in the UK. The customer dials up the service provider and the call is answered quickly by the autoanswer equipment within the service provider's premises. At this point, the network starts charging at the premium rate, which will be shared between the network and service providers. However, instead of connecting the customer straight through to the required service, the autoanswer connects to a synthesiser which synthesises the sound of ring-tone. The customer believes that the service is probably a bit congested but holds on, unaware that call-charging has begun. After a period of time, the autoanswer equipment switches the customer through to the requested service.

This is a difficult and dangerous area in which regulators play, as it can lead to unnecessary interference. There is an accusation levelled at one regulator regarding its objection to a premium rate fax delivery service: the intention was that customers dialled up the service and received the information via their fax machines. A premium rate service would be used and the cost of the call would be shared between the network and service providers. However the information was valuable and the duration of the fax call was not generally long enough to generate a call charge equal to this value. The service provider devised a way round this: fax transmission is made at a range of different speeds, depending on the quality of the line. The equipments at each end undergo a training session at the beginning of each call, in order to establish the fastest possible speed. However, in the case of the fax service, the service provider arranged that the fax modem should operate at an artificially slow speed. This increased the call duration and, hence, the cost of the call. Provided that the customer was informed of the likely total cost of the call, the author can see nothing wrong with this arrangement, but, it is alleged, this was not the view taken by the regulator. There may be more to the issue than this, but it is clear that regulators have a fine line to walk between protecting the customer and encouraging new services.

Cross-border Regulation – the European Union's Opinions

'The emergence of new services and the development of existing services are expected to expand the overall information market, providing new routes to the citizen and building on Europe's rich cultural heritage, its potential for innovation and its creative ambitions.' (European Union, Green Paper on the Regulatory Implications).

It is completely natural that the European Union has given considerable thought to issues of communications, computing and entertainment, since these are huge markets for products and services that are particularly easily transported across national boundaries, because they often travel invisibly in electronic form. The Commission sees itself on the horns of a dilemma: on the one hand, it accepts that regulatory barriers could inhibit the growth in new services but, on the other hand, it sees that existing regulation is insufficient to halt anti-competitive behaviour. One view it does support is that existing regulatory practices, which were developed for the analogue distribution environment with little convergence of the different media, are significantly inadequate. In effect, it appears to be proposing that a convergence of services, organisations and technologies requires at least a degree of convergence of regulation, although it does not accept that there should be an entirely uniform regime.

Convergence at Different Levels

The European Union considers that convergence should be seen as occurring at three different levels, as shown in Figure 8.8. It should

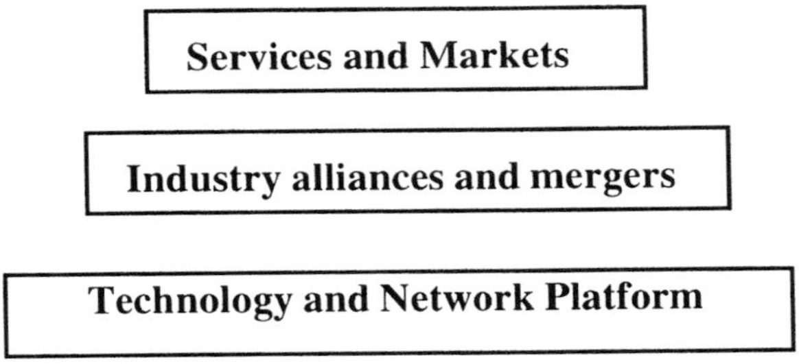

```
┌─────────────────────────────┐
│   Services and Markets      │
└─────────────────────────────┘
  ┌───────────────────────────────┐
  │ Industry alliances and mergers │
  └───────────────────────────────┘
    ┌──────────────────────────────────┐
    │ Technology and Network Platform  │
    └──────────────────────────────────┘
```

Figure 8.8 Three levels of convergence – European Commission view

not be assumed, they believe, that convergence at one level will necessarily occur at the same pace, or in the same way, at the other levels. From a technical viewpoint, this could have significant impact on service: the EU could create regulatory boundaries in different places within each of these levels, which would require companies to construct systems that allowed them to separate out their costs, billing, fair access, etc., in order to avoid charges of predatory behaviour.

Technology and Network Platforms

There are no surprises in the technology issues that the Commission consider significant:

- High-speed telecommunications networks
- Digital TV, with a consequent increase in the number of programmes and their manipulation in time or combination in personalised channels and 'pay-per-view'
- 'Multimedia data broadcasting' for downloading of computer programs and games, and direct access to the Internet from the TV set or network computer
- The Internet: the EU place particular emphasis on the Internet as 'both the symbolic and prime driver of convergence'.

The Green Paper has quite a lot to say on the subject of the Internet: it notes that, unlike broadcasting and telecommunications, it has been created by local groups of users rather than centrally planned. it is further peculiar in that it provides both publishing and private communication, and this in one-to-one, many-to-many and one-to-many modes. Thus it straddles the differing regimes of traditionally regulated broadcasting and traditionally unregulated private conversation.

The Commission is cautious about selecting the dominant 'home platform' for viewing and interacting with these services. It notes a significant shift from TV viewing to Internet surfing, quoting US sources that claim PC sales are not greater than TV sales and Web users are watching 59% less TV than average. (However, they also note that European TV watched per day is increasing slightly.) The emphasis seems to be on interactivity, either via games machines or on the Net. (Might this be seen as encouraging for interactive TV technology?)

What the Green Paper does not dwell on is the cost of the platform. This may be a critical issue in determining whether the home platform will be a Web TV, PC or networked computer.

The Commission also notes that pricing policies for telecommunication connection charges are inhibiting wider use of the Internet. In Europe, these charges are based on connection duration and distance. This is in contradistinction to Internet and TV channels whose tariffs contain a significantly larger percentage of flat-rate subscription. Introduction of regulation in this area or, perhaps, regulatory support for new entrants that offered flat-rate services, would significantly affect the profitability of traditional telecoms companies, who would need to offer competitive packages. This would impact on their call-metering and billing systems, as well as their number plans and local call area dimensioning.

The regulators are also favourably disposed towards Internet telephony and to the integration of Internet communication with the more traditional line-based telecoms and the emerging mobile telephony market. They do recognise, however, that in two-ended services such as communications there are significant quality of service issues surrounding the physical interconnection of the highly controlled telecommunications networks and the 'best-efforts' nature of the Internet. (The prospect of allowing Internet customers to have access to telecommunications switch control systems is not something that one would feel particularly easy about.)

Industry Alliances and Mergers

Regulators are particularly concerned to monitor the structure of the supply side of the market, in order to detect signs of anti-competitive practices. Table 8.1 gives a selection that the Commission have looked at.

The rationale for these horizontal mergers or alliances is the recognition that the resources and skills required to cover every part of

Table 8.1 Alliances and mergers examined by the European Commission

Rationale	Example
Increasing market power/gaining minimum efficient scale	Vebacom – Urbana Systemtechnik, Cable and Wireless Communications, Demon – Cityscape
High cost of new (digital) technologies	Canal Plus – Nethold
Uncertain demand for new services	Multimediabetriebsgesellschaft (Kirch, Bertelsmann, etc.)
Internationalisation	BT–MCI, Global One, UUNet – Unipalm Pipex
Opportunities arising from regulatory reform	MFS/Worldcom, Telenet Flanders, NYNEX/Bell Atlantic
Uncertainty of demand	Hughes Olivetti Telecom (DirecPC), @Home
Market positioning and access to new skills	Bertelsmann – AOL, BBC WorldWide – ICL, STET – IBM
Gaining control of channels to the customer	BT – BSkyB, Disney – ABC – Capital Cities
Moving into higher-margin areas of the value chain	Microsoft Network – NBC (MSNBC Internet new channel)
Stave off competition from companies in related markets	US West – Time Warner, Oracle – Sun – Netscape (Network Computer)

the value chain do not reside with any one player in today's market. The Commission accepts this but feels it must intercede to make sure that the public interest is served. It has, in fact, done just that in a number of cases where it felt that barriers to competition were being erected. One hopes that the Commission also realise that mergers or proposed mergers can fail – for example, the BT–MCI venture quoted in the table – and that this possibility is built into their plans.

Regarding specific market sectors, the Commission are concerned that monopolistic ownership of multimedia content could lead to unfair pricing, or at least a hike in the price of content rights. We might conclude again from this that system and service architectures should be capable of coping with any regulatory boundary issues. On the other hand, there is also the fear that insufficient protection of Intellectual Property Rights (IPR) might inhibit the growth of innovative new services. Bearing in mind that one of the major benefits of digital coding and electronic storage and transmission of content is that it is easier to mix, match and reuse it across a range of delivery channels, there is a clear need for technical means for protecting it and charging for it across these diverse channels. We have looked at some of these security aspects in Chapter 7 and will discuss them further in this chapter, when we look at social and ethical issues.

National versus European Issues

The European Union is not just a single entity: it is a collection of individual nations, with firm ideas about their national identities, customs, languages, legal systems and so on. As we have said, however, electronic entertainment and information can slip easily across boundaries and, increasingly, a number of cross-media issues arise. One example cited by the Commission is the case of the French election campaign: there are rules prohibiting newspapers from publishing opinion poll results in the week before voting, but nothing to stop them being made available on the Internet. This appears to be discrimination based on the delivery mechanism rather than on content. In theory, similar discrimination could arise over satellite versus cable/wire delivery, to give but one further example. Regulatory uncertainty could be a major inhibitory factor to the introduction of new services, particularly if different countries decide to submit identical transnational services to different regulatory regimes – broadcasting versus telecommunications, for example.

The general position on licensing across Europe is currently a minefield: not only does it vary between the three market sectors of telecommunications, computing and broadcast entertainment, but there is also significant variation in terms and conditions from country to country, and in the way they are applied. Some, particularly for cable TV, are even licensed at regional level (or over even smaller areas), whereas telecommunications and broadcast are at least national. Computing technology is virtually licence-free. The Commission states a view that licensing should be reduced and harmonised across the media.

It is also their view that access to networks and to content should be based on commercial rather than regulatory agreement, except where exploitation of monopoly positions is clearly evident. Yet again, the implication for technology is in costing, billing, etc.

Conditional Access

High-quality content is expensive to create and it is only right that, in most cases, it is not unreasonable to expect the user to contribute to its cost. Conditional access to this material, by subscription or pay-per-view, is a well-established practice. However, it can be open to abuse, particularly if the provider owns more than one link in the supply chain. Consider the case of the provision of TV set-top boxes; if these are provided by the owner of programme channels, it is theoretically possible that they could be biased in favour of these channels: given that digital TV can create an airspace containing hundreds of channels, the viewer will need some help in selecting which to watch. Electronic programme guides (EPGs) are browsing and navigation systems that can be used to select and present snippets of programmes (perhaps in a dozen or so small frames within the screen) that the viewer might latch onto. These EPGs tend not to be disinterested parties to the selection of programmes; they can be created by a number of critical sources, but also by programming companies themselves. A streetwise set-top box supplier-cum-programme-maker might be tempted to favour its own programmes in its own EPG. This may require regulation.

Radio Frequency Allocation

The radio spectrum is a fixed and already crowded resource. There are many new services which rely on spectrum allocation for their success. At present, the amount allocated to broadcast versus voice and data telecommunications services varies, as do the terms and conditions. For example, a broadcaster given low-cost access to spectrum might provide an information service in competition to that provided by a mobile telecommunications company that has paid a great deal for the same width of spectrum. Clearly, this ambiguity and unfairness is a further inhibitor to new developments. One possibility is to remove the restriction currently placed on a licensee to use the spectrum for a particular, specified service, with the only restriction being on safety and electromagnetic interference. The Commission would also reserve some rights because it attaches strong importance to using this scarce resource for activities generally beneficial to the European Community, particularly for social and regional cohesion and, in the case of telecommunications, of universal service.

Summary of European Regulatory Principles

The European Union's Green Paper has spelt out its views on regulation in a time of convergence. It summarises the position as follows:

1. Regulation should be limited to what is strictly necessary to achieve clearly identified objectives.
2. Future regulatory approaches should respond to the needs of users.
3. Regulatory decisions should be guided by a need for a clear and predictable framework.
4. Regulation should ensure full participation in a converged environment.
5. Independent and effective regulators will be central to a converging environment.

We must remember that Europe may propose but the whole world, particularly the USA, will decide. Nevertheless, most people would agree that these are reasonable aims, if they can be achieved. It is important to realise that technical designs and architectures must acknowledge the power and the beliefs of Regulators and how these will evolve, if we are to create systems that are flexible enough to sustain profitability.

LEGISLATION AND LITIGATION

The opening up of new markets for information and entertainment using the power and range of computing and telecommunications creates a requirement for a new look at laws and rights, not just in a national context but also globally. The new technologies generate different ways of combining different types of materials, of transmitting them much more frequently and freely across cultural boundaries and, at the same time, opening the way to new forms of piracy of people's creative talents. These are issues that require the new technology not merely to make the services possible but also to provide them in an appropriate way while protecting the natural rights of individuals.

New Media Require New Rules

Although lawyers are generally relaxed in principle about the ability of existing legislation to cope with new information networks, they do appreciate that it will generate a large number of interesting (and, to them, lucrative) individual cases.

Because the new multimedia offerings do not easily map one-to-one onto existing modes of presentation, problems are bound to arise.

Detailed law is predominately concerned with the way things are divided up at present. Take the case of 'broadcasting': BT wishes to offer a number of video-on-demand (VoD) services that it might be forbidden to provide under UK Telecommunications Acts, were they to be described as 'broadcast'. Consequently, the company argues that VoD is not a broadcast service, because each household effectively views a personal instantiation of the programme. If this argument is accepted, however, then certain difficulties arise about royalty fees: the royalty arrangements for actors and actresses have been agreed for *broadcast* performances, but if VoD is not a broadcast service, what is it, and do new agreements need to be negotiated?

Video-on-demand is also worthy of study as a good example of a service that cuts across the boundary between online information services and broadcast entertainment. In some countries with commercial broadcasting funded by advertising revenue, there are restrictions on the amount, timing and nature of the advertisements that can be shown. This is all based on serial, non-interruptable broadcasting. But the major feature of VoD is its individualised, non-serial, when-you-want-it delivery. There is no such thing as a centrally planned evening's entertainment, for instance. It is difficult to see how or why regulation of content or advertising should be achieved. To look at a related aspect, there is no 'watershed', say 9.00 p.m., after which programmes of a more explicit nature can be shown. (We shall discuss later how programmes might be labelled with advice on their content.) One technical solution is to password-protect access to such programs, so that only selected members of the household may initiate access.

Multimedia and Media Rights

Consider the problem that might face someone wishing to create a 'virtual art gallery', where people could view, online, the great paintings of the world in an attractive and instructive set of surroundings. It is easy to cut and paste the images into a Web page, together with text; it is not even particularly difficult to arrange for each view to be accompanied by appropriate music or a spoken commentary. But what is difficult is putting together all the copyrights involved: rights to the painting will be reserved by the gallery which purchased it, or perhaps the artist or the artist's estate the lent it; the commentary may be by a living author, extracted from a periodical; the music that the orchestra plays may be still under copyright to a composer; the orchestra may also have rights as well. Not only would the owners of the virtual gallery have to identify the owners of all of these rights, they would have to negotiate terms and conditions for the publishing of the material, in the new, online form. The copyright issues underlying this have all been developed in different industries that have each evolved their own history of rights and laws which, though generally consistent within the field, may not carry over to others.

At the very least, the museum owners are likely to come across one party who is going to be reluctant to accept a one-off payment for the rights. Since no-one really knows how lucrative these online services are going to be, the gallery owners will not be given an open right to generate money. They are likely to be asked for a 'performance fee', based on the number of times the service is accessed. While this is a problem to them, it might be an opportunity to the network distributor, who could create a billing system that charged not only for delivery but also a fee for the content.

The issue can become even more complex, however. Multimedia editing tools allow fragments of material from different media – text, sound, image – to be cut and combined in different ways. A certain amount of 'free' cutting has always been permitted. For example, in reviewing a book, it is permissible to print extracts; it is also possible to quote from an expert in a technical publication.

As far as images are concerned, US copyright law protects the 'derived rights' of a creator, to the recasting, transformation or adaptation of the images, provided the 'total concept and feel' is retained. It might therefore be possible to change the clothing, the expression, the voice of an existing image, and create a new situation for it, without being found guilty of copyright.

A particularly interesting variant of this is almost possible now, as a result of some of the technology described in Chapter 6: we can take the voice of a well-known actor, analyse it into sub-word fragments and use these to drive a speech synthesiser. We can also copy the image of the actor and deconstruct it into segments that can be used in an animation program. We can now create a movie in which our virtual actor moves and speaks, just as the real actor would, but does and says things that the real actor never did. Fortunately, the legal issues that surround this are outside the scope of this book!

Protection against Theft of Rights

The delivery of online content provides a greatly enhanced potential for copyright theft. This already happens with physical material such as video and cassette tapes, where a significant industry has grown up. Numerous attempts have been made to develop technical countermeasures, but with rather little success. The problem is a fundamental one: whenever content is experienced then it can, in principle, be copied, because the signal stream that is delivered to the display device (TC or computer screen, radio loudspeaker, etc.) can always be intercepted. The best that one can do is either to make this interception as difficult as possible, or to mark the copy in some way, so that its dubious origin can be made clear.

One deterrent to copyright theft is to mark the material in some way so that it can be seen to have been stolen. This can be done either overtly, and is therefore conveniently combined with a 'spoiler signal' which can degrade the quality of the stolen item, or covertly, in the form of a secret 'watermark'. This is an example of the technique known as 'steganography', the concealment of one message within another. The author developed a simple technique many years ago for watermarking printed documents in this way. Each time a copy of the document was printed off, the proportional spacing of the words was modified in such a manner that a hidden copy-number could be put onto the document. The intention was to allow illicit photocopies to be traced back to the originating version.

With images, there are a number of similar tricks that can be used. Still pictures in formats such as '.gif' can be modified by changing their colours to adjacent colours on the colour palette. We have to be careful, of course, that the watermarking is not lost if the image is subsequently passed through some other image coding scheme which does not distinguish between the two colours.

A possible technique for use in speech or music signals is to add a choice of echoes to them. Provided these are chosen carefully, there is little perceived degradation of the signal, and by switching between the echoes a rate of a few bits per second can be achieved. The decoder works by 'correlation': the signal is split into two channels and each one multiplied by a version of itself, delayed by one of the delays. The bigger signal is obtained from the correct multiplication, thus the secret bit-stream can be recovered (Figure 8.9).

SOCIAL AND ETHICAL ISSUES

Traditionally, when a group of 'freedom fighters' or 'terrorists', depending on the point of view of the writer, wish to make a symbolic

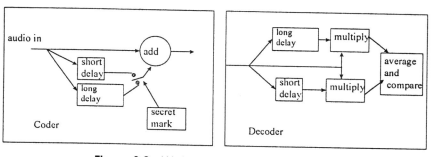

Figure 8.9 Watermarking an audio signal

gesture, they blow up the telecommunications centres and occupy the broadcasting station. Communications is a strong symbol of government and social infrastructure and it is inherently a focus for political and ethical views. It is instructive to reflect that the 'free market' approach of the computer industry and the standardised, central control of the telecommunications business are mirrored in the political vision; the convergence debate is not just about technical and marketplace realities but also about deep cultural issues. It is unlikely that these will fail to have an impact on the pace and direction of the growth of information networks. There are, for instance, issues of free speech versus privacy and decency, equality of opportunity to take advantage of new technology, as well as the rights of the creators of multimedia content and the ability to enforce them. By no means the least important is the relationship between the state and the multinational employer and the implications of globalisation on employment. We shall look at these, in terms of the technological impacts and implications.

UNIVERSALITY OF SERVICE

The computer paradigm is 'equality under the marketplace', the right for anyone to purchase equipment and, within very broad limits, do what they want with it. The free-market model has been shown to be a powerful promoter of democracy, but there are problems: for instance, the Internet community has a tendency to demand flat-rate (rather than duration-based) charging for connection to an Internet service provider, and there is an argument that says call charges should be distance-independent. Both of these demands militate against the low-wage, perhaps elderly or disabled person, living in a part of the world where it is unattractive to provide good communications access. Even something as simple as guaranteed telephone operation when there has been an electricity power supply failure (because telephone exchanges provide power from an exchange battery supply) offers a security comfort to someone living alone, with the lights out. It is a service, however, that is difficult to provide over optical fibre.

GEOGRAPHIC EQUALISATION

The official documents of the European Union often refer to 'less favoured regions', which is a euphemism for parts of Europe where unemployment and poor general conditions prevail, usually because they are isolated from centres of industry. This is something that advanced communications could, in theory, alleviate, because of their ability to substitute virtual transport of people and goods for physical transport. We have to say 'in theory', because there is a catch: the vicious circle which says no industry, so no infrastructure, so no industry. The poorest parts of the world also tend to have the poorest communications infrastructure, because market forces have not been strong enough to create investment.

Where investment has been available, and this has usually come about by partnership between government and private capital, the results can be highly beneficial. For example, because of some national and European funding, the Highlands of Scotland have access to one of the most modern networks in Europe, with ISDN digital connections available to many small companies. Two years after its introduction, at least 200 jobs had been created as a direct result. This may not seem a significant number, but, in terms of the Highlands and Islands area, where the number of inhabitants totals only 370 000, it represents a valuable increase. It may be of interest to note that the total telecoms investment was £16.25 million, representing £44 per head of population, and it might be argued that this would pay for itself in terms of reduction in unemployment benefit. The nature of the jobs created in this way is largely predictable, many being in the information industries. One company provides abstracts of medical journals, which are electronically scanned into a database and distributed over the network to the employees' home terminals, where they can be sorted and categorised, after which they are retransmitted back to the centre. From there, they are automatically transmitted to experts who create the extracts.

It is not always necessary to provide homeworking in these areas; one plus point that they usually have is the low cost of property prices. The cost of a building from which to run a computer help desk, or any other call-centre activity, can be less than one-tenth of its equivalent in an inner city.

In many cases the quality of the employees in these areas can be better than in places where there is higher employment, but it has to be said that there can be problems with regeneration in remote areas, caused by low skill levels. Here again, networked training can be put to good advantage. The Highlands and Islands initiative has shown that teaching and training standards can be improved by the introduction of videophone connections: whereas in the past it was too expensive to send teachers of minority languages out to islands around the west coast of Scotland, they now hold regular classes over the links. This way of teaching has other advantages: many island-based children had to leave their homes for five days each week to live in state boarding schools where the classes were large enough. Not only has this improved the quality of life for them and their families, but it has also reduced the education authority travel costs.

The Highlands and Islands are fortunate to possess this relatively advanced infrastructure, but what of other areas where this does not exist? Radio systems provide an answer. Very-small-aperture terminals (VSATs) are a convenient way of providing an easy to install and relatively cheap communications infrastructure. The VSAT terminal has an antenna size of about 1 metre and output power of around 1 watt. The data-rate tends to be asymmetric, as low as 19.2 Kbit/s from the terminal upwards, but as much as 512 Kbit/s from the satellite. This is quite a good set of parameters for downloading multimedia material, for example for training/educational purposes, or for data files for teleworkers to process, with the slower uplink providing an interactive e-mail or other short message channel. In fact, VSAT technology is still rather expensive for community rather than business purposes (its current main use is for returning retail point-of-sale information, or for fleet and warehouse management), but the cost of the radio frequency components is rapidly reducing and will soon make it a more attractive proposition. it is also being introduced for emergency back-up service when tactical communications are required in remote areas. Increasingly, cellular phone systems are providing coverage in quite remote areas and, with the introduction of the GSM digital telephony standards, this can be used for data services as well as voice. The communication mode of choice on most of these transmission systems appears to be the Internet, whose forgiving nature allows contact with the world to be maintained.

It is possible to be geographically disadvantaged, not because of being in a remote area, but rather because one is on the move. There are a significant number of people who necessarily lead nomadic lives: circus and fairground workers, people on barges and itinerant farm-workers who follow the harvests. Often their children suffer serious disruption to their education, which cannot be easily fitted into their lifestyle. One novel solution currently undergoing trials is to provide them with a low-cost learning environment which integrates local multimedia with low-bandwidth interactive support. The educational material is provided as disk-based multimedia, but the learning units on the disk are barred or made available by the tutor, who communicates with the learner's workstation via an Internet connection delivered over GSM digital cellular telephony. The student's progress through the material is automatically transmitted to the tutor and the tutor and pupil can send each other short messages when necessary.

DISABILITY

The messages regarding the impact of intelligent communications on disabled people are rather mixed. As the world moves more towards a communications and computing-based society, we have a duty to make sure that the interface it presents is usable by everyone. We should not leave people behind at any stage of its evolution. Unfortunately, physical or mental disability is often associated with another disadvantage – poverty, or at least a below-average income. This makes disabled people into a less attractive market if purely commercial motives are to be pursued. In practice, it certainly means that products for people with disabilities will seldom benefit from economies of scale. It is more expensive, for example, to produce a large-sized keypad for a telephone intended for use by people with motor difficulties or problems in seeing clearly. It is extremely unlikely that low-cost specially designed computers can be produced to cope with most issues of disability. What is required is probably ingenuity in applying existing products.

Problems with the Internet

One place where these issues are currently very obvious is the case of the Internet and the World Wide Web. The ability to chat and generally keep in touch with the test of the world, via the Net, has been one of its very positive aspects. However, currently we are going through a stage of development which is less beneficial.

When most of the material available online consisted of text, people with sight difficulties could get round this by means of tactile readers which converted the next to a pattern of raised pins, or by speech synthesisers. Increasingly, however, Web 'pages' consist of a number of separate frames, containing large numbers of hyperlinks, and a great deal of image material. All this lies well beyond the capability of available tactile output and the alternative of speech output is a message that is often too complex to understand.

It would seem that what is required is for the 'raw' page to be submitted to some form of information restructuring process, that can summarise or place in a hierarchy the raw information, in a manner that can be displayed on the output device. One way might be to insert 'meta-tagging', as described in Chapter 5, information that would be used to create this hierarchy, according to some set of standards. The meta-tagging would be done either by the originator or by agents of organisations representing the disability group. Sadly, it is unlikely that many originators will care to do this and the process becomes time consuming and possibly expensive for the caring agency. It is possible that the application of some artificial intelligence techniques to this mark-up task might go some way to alleviating this problem.

Issues with Public Terminals

Today in banking, possibly quite soon in retailing and public services, many counters staffed with assistants will give way to terminals. These pose a number of problems for elderly and disabled people. A number of studies have been carried out into the issues; privacy emerges as a major problem. Old people are naturally concerned that their affairs such as the state of their bank account might be more visible to others than to them. Spoken interfaces to blind people are unlikely to be acceptable if they shout out transaction details. It may be that headsets with cable or inductive couplers need to be provided.

Blind or partly sighted people obviously have trouble navigating around the streets. Here, there have been positive developments involving the integration of Global Positioning Satellite (GPS) technology with geographic information systems. As explained in the section on transportation, GPS an give a 'fix' on position, accurate to a few metres. Body-borne GPS receivers are now available for a few hundred pounds and are also being incorporated into other equipment such as cellphones and personal organisers. Integrating GPS with geographic information systems, such as digital maps, can make it easy to calculate where you are, where you want to be, and how to get there. Trials of such systems suggest that users would prefer to plan a route at home first, being able to study a map using touch and spoken information. Then, carrying a portable unit, the user can be guided *en route* through headphones similar to 'Walkman' technology.

DECENCY

No sooner than someone has invented a new way to communicate, someone else will find a way to use it in an offensive manner. This has been proven time and time again, nowhere more so than on the World Wide Web, which provides global access and minimal level of publishing cost or skill. One particular problem occurs precisely because of the global nature of information networks: on one hand, the creator of offensive material may be free from prosecution on the other side of the world; on the other hand, he or she may not be creating an offence in their own territory, although it is deemed so elsewhere. One recent case in the USA illustrates that last point precisely: a web author published some online material that was not offensive in his own state, but was successfully prosecuted because it was received in another where it was considered obscene. Although the judgement was later reversed in a higher court, it illustrates the dilemmas that arise when globalisation meets localised laws. A further problem arises when new media cross the boundaries of the old: many societies treat the printed word differently (usually more leniently) than films, TV or radio. In the last case, how would we treat someone who published a text site, but with embedded instructions that allowed someone to access it via a speech synthesiser?

Another example of a cross-border problem occurs with satellite broadcasting. Who is to be held responsible for the acceptability of the content? Is it to be the company who produces the material and first broadcasts it, or is it the responsibility of the operators of the last satellite ground station? Legal opinion differs from country to country, but the issue is global.

One approach to the problem is to adopt the policy of 'buyer beware', but to provide them with tools that make it difficult to receive inappropriate material by accident. The Platform for Internet Content

Selection (PICS) initiative was set up in August 1995 by representatives from 23 companies and organisations gathered under the auspices of MIT's World Wide Web Consortium to discuss the need for content labelling. The idea behind PICS is to provide a way whereby the producers of content and third parties, such as magazines, consumer organisations and industry watchdogs, can insert information onto a supply of World Wide Web content, using standard protocols that can be read and acted upon automatically by computers, to provide local screening out of offensive material (and, indeed, the selection of desirable content) without introducing global censorship.

PICS is concerned to establish only the conventions for label formats and distribution methods, not the label vocabulary nor who should police it. The emphasis is on freedom of choice. Although labels can be created in line with, for example, movie rating categories, 'PG' and the like, it accepts that personal standards and attitude must be expected to vary, particularly in a global context. In any case, publishers can get round these classifications by publishing in countries where standards are different. This is not simply a matter of *overall* laxity: some social groups are very anti-violence but tolerant of explicit sexual material, while in other groups the reverse is true. Although there were a number of such rating schemes in existence, PICS attempts to standardise the approach so that it is vendor independent and the labelling can be done by any number of third parties. Indeed, any number of labelling organisations can mark up a page at any time.

In Table 8.2, we give an example of a PICS label. If one were to visit the labelling service site, 'http://www.here.co.uk', one would see a similar style of text, which described the way that the rating was composed.

Table 8.2   Example of a PICS label

(PICS-1.0 'http://www.here.co.uk' labels 　on '1998.01.01' 　until '1999.0101'	• First the label identifies the URL of the labelling service and gives beginning and end dates within which the labelling is valid
for 'http://www.newtv.uk.co/film.html' by 'Bill Whyte'	• Then it gives the URL of the page that is labelled and, optionally, who labelled it
ratings 1 2 s 1 v 4))	• Finally, it gives the rating (language = 1, sexually explicit scenes = 2, violence = 4

As we said, the labelling service can do what it likes. It does not need to restrict itself to moral judgements. In fact, it does not need to rate pages from a moral point of view at all. It could choose to rate recipe pages on how expensive it thought the recipe was, or even the spiciness of the food. The user of the service would set the filter software on her machine to select on any rating basis that she required. The rating service she chose would depend on the credibility to her of the service: vegetarians would be unlikely to use a food-rating service provided by the beef marketing association; Baptists might choose a film rating service offered by an organisation with religious connections, whereas humanists would probably not. It is even possible to include a time of day filter on the pages: a PICS filter on a server in a school could ensure that, for the duration of a self-learning period, pupils could only access material relevant to what they were supposed to be studying.

Regarding the management, distribution and control of labels, two questions might be asked: how do we make sure that the PICS label covers the material to which it is attached, and how do we know that the label has been applied by the claimed labelling agency? One simple way is to access the information via a reputable service provider, who will take steps to ensure that the material on the server is not provided under false pretences. An even stronger guarantee can be got through two other features of PICS: one is the ability to apply a 'watermark' to the material, for example in the form of a hash function, as explained in Chapter 7, which affixes an encrypted summary of the page; the second security feature is the ability to include a digital signature (again see Chapter 7) of the labelling authority.

The major advantage of PICS would appear to be in its role in permitting selective censorship at point of receipt. There does not appear to be any other way to achieve this. Text searches can be carried out to block off pages that appear to contain offensive words, but it is much more difficult to deal with images in this way. In any case, the creation of 'stopper' lists of unacceptable terms brings us back to the problem of subjectivity and the possibility of misuse. (In this latter regard, see the frame on page 294, for a cautionary example.)

There are a number of ways the PICS labels can be applied and filtered, as shown in Figure 8.10. For instance, the originator of the page can insert the label as a 'meta-tag' (see Chapter 7). Then the user's browser can read these and accept or refuse them according to profiles set up in the browser.

Alternatively an intermediate server, possibly operated by an

THE STAKEHOLDERS

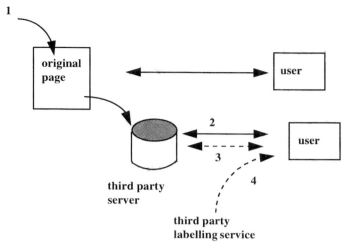

Figure 8.10 Applying PICS

independent third party, can be asked by the user's browser to supply labels along with the document.

An extension of this is for the intermediate server to supply its users with pages it has prefiltered, to save them the problem of doing so.

The client could make use of an independent labelling bureau, which only provided labels. Essentially this company is creating selective filtering, based on value judgements of assessors, critics, etc. These third-party labels are 'seals of approval;' and content producers might pay to have their products assessed by leading brands of labelling authorities. (Hypothetically, the seals now granted to certain manufacturers could be paralleled by online content labelled with 'by Royal appointment', for example!).

It is even possible to imagine a democratic labelling service where users who access sites individually rate the material, from which an average set of data could be used to generate a reduced set.

PRIVACY

There is general feeling that people ought to be able to converse innocently, in private, without fear of being overheard by their neighbours, the state and so on. There is also a belief held by many that the police and other forces of law and order need to do all they can to avoid criminal conspiracy. It is sometimes difficult to reconcile both of

> **FILTH IN THE SCIENCE MUSEUM**
>
> It is claimed that, some 20 years ago, a demonstration was laid on in the Science Museum, to show the principles of data communication to children. People were allowed free access to type text on one teleprinter and see it printed out on another. Children loved this, because they were able to transmit rude words to their friends at the distant terminal. To stop this, a computer was interposed between the two terminals and programmed with a list of banned words, which it suppressed. Inevitably, an ingenious child, correctly deducing that the terminals were not merely capable of end-to-end transmission but also of accessing the computer, requested the latter, via the terminal, to list the contents of a file, whose title, 'swear', 'oath' or whatever, was easy to guess. The receiving terminal thus began to print out all and only those words considered by the august members of the Science Museum to be unsuitable.

these viewpoints, and this is a particular issue with the development of global mass communication, together with methods of protecting its privacy or automatically intercepting it and analysing its contents.

In his novel *Nineteen Eighty-four*, George Orwell forsaw a society in which the mass of the population was spied upon constantly, via television cameras and microphones in every house. What Orwell conveniently ignored was that for everyone being watched, there would need to be a watcher. This is, of course, only true where a human is required to do the watching. In Chapter 6 we described a surveillance TV application involving the remote, automated surveillance of public areas such as car parks in order to detect anyone behaving in a 'suspicious' manner. Technology will make this sort of intelligent surveillance much more common, or at least possible.

An unanswerable question is 'How much right to privacy of one's communications does one have?'. It is unanswerable because granting unrestrained freedom to one individual can often lead to restricting the freedom and rights of others: is a terrorist entitled to plan, in private, an outrage that might maim innocent victims? Is a government entitled to eavesdrop on a militant activist whose sole intent is a bloodless coup? It is also unanswerable in practice, because governments will spy and criminals and terrorists will conspire, whatever the current public opinion, if they can get away with it. Instead, what we can ask is how changes in technology will influence the balance between privacy and listening in, and look for disproportionate changes that require us to form a value judgement.

INTERCEPTION OF COMMUNICATIONS

In the language of security there are a number of words that are used to describe the unauthorised acquisition of someone's communications: 'tapping' is used to mean making a connection to a line; 'bugging' is the fitting of a device which can pick up signals and send them back to a monitoring point. We shall be particularly interested in the picking up of signals from telecommunications or data networks, but rather than become involved in the specialist's debate over names, we shall simply refer to any overheard messages as 'interception'.

In the past, the interception of telephone or telegraph signals was quite simple, at least in principle. One connected a pair of wires to the legitimate pair, to 'tap off'' a part of the signal, and led the wires away to a safe listening point. Radio interception was even simpler: all you did was erect an aerial and tune in to the wanted broadcast. Of course, unless you knew where to look, finding the right frequency took time, and the enemy employed a frequency-hopping technique, identical in principle to that used by the frequency-hopping LANs mentioned in Chapter 5, although with hop times limited to tens of seconds as the operator had physically to change the tuning coils. Today, it is not just secure radio systems that employ frequency hopping; as we saw in Chapter 4, it is the normal mode of operation of cellular radio systems when they move about, and when they next make a call they are allocated the first available frequency. However, this in itself does not provide security against interception (although one might hypothesise that it would make routine surveillance by a government agency a bit more elaborate and expensive) because the instructions regarding the frequency changes are carried in the control channel, which can also be intercepted. Indeed, there have been a number of instances of freelance, opportunistic interception of cellular conversations, widely reported over the last few years. The speech in earlier cellular radio systems, was transmitted as an analogue signal; for quite fundamental reasons, there is no way that this analogue channel can be securely encrypted. Only with the introduction of GSM's digital coding of speech has it begun to be possible to provide really secure cellular communication, using the digital encryption techniques described in Chapter 7.

THE LAW ENFORCEMENT VIEWPOINT

The law enforcement and security services' viewpoint, that the availability of secure digital communication is not necessarily a good thing, is not without merit. Although they are naturally reluctant to reveal too much detail, it is obvious that they rely on intercepted communications for a significant part of their success in foiling crimes and offences against the state. Undoubtedly, the availability of 'hard' encryption makes their job much more difficult. However, they have been greeted with little cooperation from the rest of the community in their efforts to redress the balance. What they would like is the ability to break the encoded material with as little effort as possible. One way might be to control the use of certain encryption techniques, and this has been tried in the USA, where strong public-key systems are considered to be weapons of war and thus not exportable (or even usable within the country, under some circumstances) without a licence (which in practice would not be granted for the criminal purpose). Unfortunately for the US government, this is proving exceedingly difficult to enforce, as the techniques are based on software rather than hardware and can easily be transmitted abroad. Moreover, individuals in the USA have pointed out that a strong indigenous encryption capability is growing up outside the USA, fuelled by the ban, which is thus beginning to look counterproductive to its originators.

The next move by the authorities is also running up against strong opposition: this is the so-called 'key-escrow' principle, the proposed requirement that users of cryptographic systems deposit the decrypt key with a 'trusted third party' for 'safe-keeping' (actually, in order that the authorities, armed with the necessary legal injunction equivalent to a warrant to tap a telephone, can gain access to it). Some European governments, as well as that of the USA, have been flirting with this possibility, but it will be hard fought: the European Union itself does not appear to favour legislation and, it is reported, the British banks have violently objected to any involvement on their behalf.

The banking issue is significant: it is not only encrypted criminal conversations that are involved; there is also the issue of electronic money. Payment in cash has always been the preference of criminals, because cash leaves little or nothing of an audit trail. Unfortunately for them, it is also difficult to transport because being found in possession of a large amount instantly attracts suspicion. Now, with the arrival of electronic money, there is a big issue as to whether the money should carry its history with it, in the form of an encrypted, electronic transaction record, and, if so, who should have access to that record.

EMPLOYMENT AND CITIZENSHIP

When new technologies are introduced, they often lead to dramatic changes in the most basic rights of individuals: the Agrarian Revolution led to loss of land and the right to farm it (highly inefficiently) for the majority of the population, leading to loss of independence and the rise of the 'wage slave'. The Industrial Revolution sucked the disenfranchised landless into the towns and radically transformed the relationship between employer and worker. Exploitation and opportunity shifted their centres of gravity and groups of new rich and new poor arose. At the beginning of such revolutions, government and social values tend to trail along behind the technological changes and are frequently incapable of working rationally to alleviate distress or unrestrained greed. This is likely to happen again, with the rise of a society based on the power of global information ownership and deployment. It is unlikely that the move to a global information society will happen without attempts to control it; already, in some less 'open' societies, the state is trying to take control of the influx of foreign information and personal access to it; there is some talk of the creation of 'National Intranets' which will censor inappropriate messages and be the only access platform for the citizens. But, in these cases, how do they control satellite broadcasting, provided the broadcaster wishes to give air-time to alternative views?

Work can cross over boundaries, as can the sale of products. All this presents a major change to the way we think about the duties of the nation-states.

BUSINESS AND THE NATION-STATE

If asked, people in western Europe and the USA would subscribe to the belief that they 'traditionally' considered themselves to be citizens of a particular nation-state that represented the land-mass on which they lived and where they shared a cultural identity and government with the people round about them. Most would probably be surprised to learn that the nation-state is a concept that is barely 200 years old (see, for example, Eric Hobsbawm, *Nations and Nationalism since 1780*) and one which, moreover, has never been a worldwide concept at any time. Our stateship and the obligations and responsibilities attached thereto may well be more vulnerable than we think. They comprise a vast range of things, including:

- taxation
- freedoms and restraints under law
- official language
- social security, healthcare and education policy
- employment regulation
- representation.

These and others can be affected significantly by the effects of global communications.

TAXATION IN GLOBAL INFORMATION TRADING

In most countries, whenever we buy goods over the counter we incur some form of taxation, usually on a differential scale depending on the nature of the goods. If we try to avoid this tax by buying our goods in another country, we shall be caught out by customs duty imposed at the point of import. If, for example, we buy a CD-ROM containing computer software in the UK we pay value added tax; if we wish to import more than the duty-free concession of CD-ROMs, we shall be taxed as well. But suppose we download the same software over the Internet; in this case we pay no tax (although the vendor may in its country). Even if governments decided that they wanted to tax such items, they could not. It would be very easy to encrypt the software so that it was indistinguishable from other, non-taxable date. One, apparently serious, suggestion, which has been greeted with incredulity or outrage by the online community, is for a 'bit-tax' that does not differentiate between valuable and valueless data (in the sense that the former has a purchase price) but merely charges per bit, irrespective of what the bit represents. It is difficult to see how such a 'landing charge' could act as a sound basis for taxation. How could it be measured, for example? What would happen about satellite or terrestrial, wireless broadcasts? What would happen if a new development in data compression halved the revenue overnight? But if the data were not to be taxed, then governments could lose a considerable amount of money and hence their power and ability to influence the lives of their citizens. The latter might not lose out; if the tax overhead was reduced, sales would increase and wealth would be generated in that way. However, the wealth would have shifted from governments and their citizens towards multinational companies and their employees.

PEOPLE – A COST OR A BENEFIT?

Although improved technology can reduce the cost of a wide range of products and thereby genuinely create wealth, if we are to be honest we also have to ask a delicate question – does everyone profit? It is a delicate question because there are a range of possible answers that include some very bleak scenarios for many people. Already, the availability of transglobal communications means that companies can, for instance, acquire the services of systems analysts in India at a wage cost less than half that of the UK, and they are usually more highly qualified. Data entry clerks are available in China for $2 per day.

This may be good news in the countries that acquire the new jobs, but very bad for those that lose them. However, it is not an issue that, if ignored, will go away, and it needs to be recognised that some skills will be commoditised because distance is no longer a barrier.

We also need to look at how fully automated processing of information will continue to affect the jobs market. We have noted the dramatic growth in call-centres, where tele-sales, help-desk and other functions are replacing local access points. The operators of these central services are very often low-skilled operatives – wages of less than £5 per hour are quite common and it is manifestly obvious that the skill levels are commensurably low. Even at these wages, the low-skill end of the call-centre market must be a target for full automation, as it has been calculated that input by customers using terminals integrated directly into a company's computer system is an order of magnitude cheaper than access via a human operator. In the last few decades, unskilled manual workers were abandoned by industry, in favour of machines; in the next decades, the same thing is likely to happen to the unskilled information worker.

9

Applications

In preceding chapters we have concentrated on technology, only occasionally referring to its applications, in order to maintain a sense of business reality. In this chapter we shall reverse the process and concentrate on application scenarios, only mentioning briefly the technology required to support them. In the introduction, we discussed a number of sizeable market opportunities and some of these will be explored in greater detail, together with some general themes such as 'virtual businesses' and some areas such as healthcare, where there are significant social implications.

TRANSPORT TELEMATICS

In Chapter 1, we described the claims frequently made about the costs incurred by traffic congestion and other travel problems. A sum of £20–£30 billion per annum was quoted for the UK (based on CEST and CBI surveys) and a figure an order of magnitude greater for the USA. These are huge sums and it is not surprising that governments and businesses have been interested in ways to reduce them.

The simple answer is selective taxation, 'road tolls', intended to move people away from congested areas and onto more efficient means of transport. But gathering tolls is not an easy business and, in any case, how do we determine where the congestion is and then select a better alternative?

Paying for Public Transport

Here new technology can help. For example, many experiments are going ahead with smartcard technology as a way of making it easier for people to pay for public transport. In Bologna, Italy, for instance, smartcards are being trialled on buses and trains; smartcard readers make it easier to carry out on-board 'ticket' inspection and 'ticket' validators are to be installed at stations. This not only assists with ticketing but can also be integrated with other payment methods, allows continuous monitoring of passenger flows and can be easily configured for multimodal (e.g. bus and train) interchange loyalty schemes. Similar experiments in Norway demonstrated that passengers' perceptions of the payment mechanism were considerably more favourable than those for conventional methods.

Buying your ticket (that is, programming your smartcard) could be done from a multimedia terminal that will also allow you to plan your journey, by displaying the standard routes and also giving online updates regarding the hold-ups and the timing of the buses and the times of interchange with trains, etc. In theory, it is possible that your integrated journey could include an airline ticket and the facility to notify the airline automatically of your current position, perhaps held up by a road accident. Although you arrived late for your scheduled flight, an alternative might have been found, or at least a no-penalty clause might be available, if this was part of the integrated ticket deal.

Road Transport

But despite the improved integration of public services, you sometimes need to travel by car. In future you may then discover that some routes will not be open to you; they may be permanently closed except to public transport or they may be flexibly regulated to meet time-of-day flows or exceptional circumstances caused by emergencies. It is even been suggested that air-quality measurements and prediction be built into the control, so that lanes or whole sections of road be closed off except to non-polluting types of vehicles.

The route control decisions would be transmitted to large overhead signs and would also be transmitted by radio to in-vehicle route planners which would recalculate the correct way to navigate round the problem. These signs could also be useful in the event of a serious disruption such as an accident and, because the underlying data is shared across the various modes of transport, could suggest a route to the station and a train to catch (Figure 9.1)!

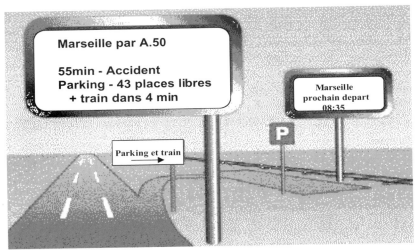

Figure 9.1 Travel information sign (reproduced by permission of EU 'STRADIVARIUS' project)

Road Tolls

None of these measures will avert the apparently inevitable progression towards road tolls. Tolls are, of course, collected today but the existing methods, largely involving cash and manual intervention, are not likely to scale up to meet the demand. The answer is something like a smartcard, except that it will probably have to be a card which can operate without contact, over a distance of a few metres. Cards such as these are still not fully standardised, but schemes such as the Gaudi project in Marseilles have been operating since 1994 with cards that plug into an in-car 'wallet' which transmits a signal to the paystation. An alternative is to use a transponder card – this is irradiated by a radio frequency field from the paystation and, in turn, 'reflects' this signal back with its own information impressed on it. Transponder cards have been proven to operate from vehicles travelling at full speed.

Ubiquitous Smartcard?

Given the introduction of smartcards for one aspect of transport telematics, we can easily see how their use could be extended to cover a wide range of applications that involve some form of data networking. We mentioned their use as payment/ticket devices on public transport and for validation and payment of road tolls – a link-up with electronic money services, as mentioned in Chapter 7.

They might also be used for providing access (and charging) to car parks, perhaps also including a loyalty rebate for discounted shopping in a corresponding shopping mall. They can provide access to airport lounge facilities and act as the prebooked ticket for rapid check-in. They could include goods-manifest details for customs clearance and, as a 'spy-in-the-cab', act as a driving licence, vehicle registration document and even the personalised component of a tachograph.

In fact, they have so many potential uses and so many potential stakeholders that the issues surrounding smartcards may be concerned with too many potential owners of the brand, rather than too few.

Vehicle Convoys

One aspect of road transport that we are unlikely to see is the heterogeneous automatic road convoy: a mixed train of cars, vans and lorries that you join in a conventional way and then simply switch over to automatic pilot. There are two reasons why this dream of mass transport control is unlikely: firstly, it would suffer from the inevitable problem of moving in convoy, one that afflicts ships doing so in wartime: the performance of the convoy is determined by the lowest common denominator of the performance of them all. Unless we are prepared to have considerably greater standardisation of our vehicles than we have ever shown a taste for, or we intend to travel at very low speeds, then this would be impossible. Secondly, it is actually very difficult to design fail-safe systems that can couple a string of vehicles together, particularly if they have heterogeneous transmission and braking systems.

Nevertheless, if we restrict ourselves to vehicles of relatively similar types, say large trucks, then it is feasible. Electronic 'towbars', operating with gigahertz radio links to send control signals between vehicles, are being trialled. The following vehicle judges its position relative to the leader by visual means: the leader may have a large barcode or a set of infra-red lamps that can easily be sensed by intelligent image recognisers.

It is easy to see that vehicles linked together in this manner and equipped with radio links to traffic control equipment could be advised on the optimum speed and route for their journey. They could, for instance, be guided into special lanes set up dynamically for their exclusive use.

Knowing Where You Are

In order to work out the best way to get somewhere, you have to know where you are in the first place. Position location is becoming much easier, with the development of radio positioning equipment; the Global Positioning Satellite (GPS) system is well known and equipment is now widely available. GPS works on the traditional positioning strategy of triangulation; signals from a number of medium Earth orbit satellites, typically five, are compared to estimate their relative time of arrival. Given their known altitude above the Earth – and the shape of the Earth – it is possible to calculate position to an accuracy of around a metre.

Although GPS is usually very effective, it does have occasional problems. The author was involved in a tracking system for locating yachts taking part in a major round-the-world race. The system performed perfectly for tens of thousands of miles, but one of the vessels disappeared completely in a New Zealand harbour, screened completely by a five-storey high, corrugated warehouse! There will always be obstructions of this nature. Quite recently, a rather worrying claim has been made that GPS is quite susceptible to 'jamming' from simple radio transmitters. Given the rise of terrorist activity in recent years, there could be the real and rather frightening prospect of major and dangerous chaos created by disruption of large-scale systems depending solely on satellite systems.

An alternative, or supplementary, solution is to provide the vehicle with some form of on-board compass. This could be an inertial guidance system using gyroscopes, as deployed in aircraft. A simpler, and probably cheaper, scheme is to take advantage of the fact that road vehicles, unlike aircraft, are always in contact with the ground; therefore, the motions of their wheels – acceleration, turning – give quite accurate measures of movement over the ground. Navigation systems using these inertial measures, combined with digital maps and intelligent software that constantly corrects for any errors they might make, constantly reposition the vehicle onto 'landmark points' such as roads, corners, etc. Absolute corrections can also be made from time to time from periodic GPS measurements or roadside beacons.

Transport Policy and Markets

It should be obvious from what has been said that improvements to transport performance are possible and, in most cases, do not rely on major breakthroughs in technology. There are, however, a number of issues that must be tackled before major gains can be achieved. First and foremost is the need to drive an integrated telematics solution from an integrated transportation policy. This is an issue for governments. In some cases, particularly in moving people onto public transport, local solutions can be provided, but, if vehicle guidance systems are to be really effective, there needs to be a critical mass of national policies sufficient to encourage vehicle manufacturers to produce systems that can be integrated into the infrastructure.

Given an integrated policy, an integrated technological solution can be put in place beneath it. Measurement of road traffic density can be used to program traffic lanes and signals; public transport can be scheduled and its arrival times notified to users of other transport nodes; variable pricing can be instantiated; and users with in-car mobile communication equipment can fire off intelligent agents to find the latest information on routes and times of arrival, and to book car parking spaces at their destination, or at railway or bus stations if they decide to change their mode of travel.

The communications and computing industries will provide the infrastructure, but they must bear in mind that the 'driving force', literally as well as figuratively, will be the consumer industry, in the form of motor manufacturers. The principal computing and communicating terminal in the car, bus or train will not be a telephone or a personal computer; it will be a vehicle accessory, appropriate to the styling, operating environment and price of that market. There are major issues of reliability, interface design and robustness.

RETAILING

In all of the developed world, shopping is one of the major industries and the supply of goods to end-customers is 'retailing'. There is much more to retailing than just 'shopping': retailing represents one of the most complex value chains between supplier and customer, as we can see from Figure 9.2.

There are intensive communication paths between each of the components in the chain, including the following.

- *Suppliers and HQs.* Increasingly this is moving towards automated ordering, based on electronic data interchange or EDI (see frame, Chapter 2) designed around the retailer's processes. Although suppliers are sometimes reluctant, they are told that they must adopt electronic trading if they are to get the business. As a compensation, in some cases they are allowed access to the retailer's sales forecasting databases. This gives them a better opportunity to plan their manufacturing volumes and, very importantly, builds trust between them and the retailer.
- *HQ, warehousing and distribution.* The holding of excess, perhaps perishable, stock is an expensive and wasteful business. The automation of stock control and the transmission of information between the warehouse and HQ allows this to be minimised. Predictions at a regional level, for example weather forecast data, can be used to ensure that warm-weather food is available at the right place. Fleet management, in a time when thrice-daily store replenishment is common, requires a rapid flow of date to the right place and the ability to integrate in-cab information with store delivery times and the sort of route-planning systems we discussed in the transport telematics section. We must not forget the bane of the retailing and distribution trades – 'shrinkage' (theft) – which can amount to as much as 10% of profits. The industry spends more than £150 million every years on anti-theft equipment, but the annual cost of retail crime is around £1.5 billion. Tighter control of money transactions through the online messages from tills and security tagging of

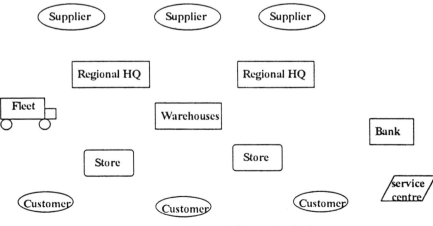

Figure 9.2 The retail supply chain

goods, together with increasingly higher quality from networked surveillance TV cameras, are some of the countermeasures used.
- *Head Offices and stores.* The availability of reliable and affordable computing and communications is a significant factor in the observed tendency for HQs to demonstrate more and more hands-on control of stores. Surprisingly, perhaps, there are still quite a number of food supermarkets which stock their shelves on the basis of someone walking around the observing what has been sold, but electronic point-of-sale data collection is becoming the norm in many retail sectors, allowing planners in headquarters to plan stock movements and to observe trends much more quickly.
- *HQ to customer.* Headquarters are also becoming much more actively involved directly with their customers, through the use of better information from loyalty cards that allow them to profile spending behaviour down to the individual. Online information services using telephony or the Internet are showing significant growth.

'Shopping'

Shopping is, however, the heart of retailing, and deserves to be looked at individually. It is a complex process and one that can be instantiated in a number of ways, but Figure 9.3 describes some of its key components.

- A 'window' (or a catalogue or a TV programme) is required in order to display the goods.
- The display must be *'seductive'* – shopping is an emotional experience and requires that the goods and their surroundings appeal in that way to the customer.
- Seduction is not enough: we must support the intellectual processes of the would-be purchaser by supplying information on the price, range of size, colour, etc., perhaps with details of how it is used.
- All this would be pointless were we not also to provide some way whereby the customer could purchase the goods, or at least enquire for further details. We need a 'transaction' capability. This is also required for post-sales activities such as servicing requests and also for outbound marketing.
- Finally, a fulfilment channel is required, in order to deliver the goods to the customer.

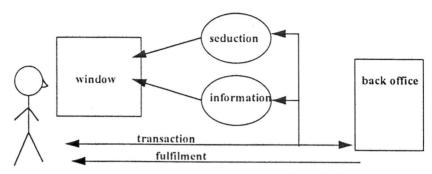

Figure 9.3 Key components of 'shopping'

We are quite familiar with these concepts in terms of 'real' shops and also in the case of catalogue shopping, which currently accounts for nearly 5% of purchases in the UK and the USA (exclusive of auto sales). In this catalogue, it is the coloured photographs of the goods, set in attractive or exotic locations, perhaps modelled by flawless models that provide the seductive component; text provides the information and a postal address or a low calling rate or freephone number provides the means of transaction.

Home Shopping

Today, catalogue shopping provides most of the reality behind the term 'home shopping'. It relies on a considerable amount of computer–telephony integration for its success: most transactions are carried out by telephone and backed up by online databases. Virtual private networks allow calls to any one number to be distributed, according to load and time of day, to any number of call-centres around the country (or even to other parts of the world), in some cases on a 24-hour, 7-day a week basis, enabling 'instant gratification' to be achieved. Experiments have begun with voice response systems (Chapter 6) which recognise spoken requests within a tightly controlled dialogue. Accuracy is not perfect, and it may not become so in the foreseeable future, but clever design can lead the unsuccessful user to a real operator without alarm or annoyance. In some cases the telephone has been replaced by Internet ordering, whereby the customer completes a simple order form.

There is an obvious advantage to the retailer with Internet, rather than telephone, ordering: no need to employ a human assistant. Indeed, some industry calculations suggest that the whole order-handling process can be up to seven times cheaper. Against this is the relatively small population of home Internet users, but the general arguments regarding technology fear appear to be groundless. It seems that people are quite prepared to use reasonably designed interfaces when they are motivated to make a purchase, and some trials have revealed that people prefer 'talking' to a computer than to a human, particularly when, for example, discussing intimate details of their personal finances.

We should also note that a home shopping company knows something for sure that a high-street retailer can only guess at: the likes and dislikes of individual customers. Each transaction between a home shopping company and a customer is one-to-one. Similarly, in an electronic shopping activity, whether it involves ordering at home on a special terminal or in a shop-in-a-shop scenario involving a multimedia kiosk, each successful transaction can be recorded and analysed against the customer's records. Even more than that, by capturing key-strokes, not just from successful transactions but also those concerned with examining and rejecting goods in the electronic catalogue, a retailer will know which lines need to be abandoned as well as those that sell well.

But how do we motivate people to make a purchase? Again, there are some lessons to be learned from catalogue shopping. The sales volumes across the decades reveal a number of peaks; anecdotal evidence (retail is a business heavily based on legend) claims that some of these are simply due to changes in the catalogue – the introduction of a new seduction component: line drawings, to supplement the early test-only versions, then black-and-white photography, then colour. Unfortunately for the retailer, the effect soon wears off, and new methods of seducing the customer are required. This is a lesson which needs to be carried over into electronic means of delivery. Even today, the image quality from low bandwidth services may be adequate for 'serious' tasks such as cooperative working, but, apart from a brief novelty element, it may not be sufficient to woo the customer.

Multimedia Kiosks

This fact has been taken into account in the design of some early schemes for 'shopless' shopping, in the form of multimedia kiosks that can be located in any public or semi-public place (Figure 9.4). Kiosks are commonly built out of a number of standard components, as shown in Figure 9.5. They illustrate all the aspects of the simple

Figure 9.4 A multimedia kiosk (reproduced by permission of BT Labs)

model described earlier: the *window* is obviously the screen, which is a standard computer monitor, usually with a touch screen.

Kiosks can even provide a degree of direct *fulfilment*: they can be equipped with printers so that they can issue tickets or vouchers once the goods have been chosen from the screen. In fact, the printer is often the most troublesome component, liable to jam and required to be reloaded with paper. As we mentioned earlier in this chapter, in connection with transport telematics, the use of smartcards in place of paper tickets would enable the use of a much more reliable card interface on the kiosk.

There will be a communications facility – simple analogue telephony or basic-rate ISDN – that allows the user to *transact* with the retailer. If ISDN is used, then a low-definition video connection can be

APPLICATIONS

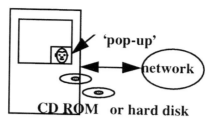

Figure 9.5 Kiosk components

made. This can provide a 'pop-up' assistant, who could be summoned on request to deal with difficult transactions. A variation on this is to use a camera at the kiosk end, to provide a security feature, such as customer identity for banking applications.

However, the provision of a *seduction* component is not necessarily provided by the network. It is not really possible to provide high-quality moving images over basic-rate ISDN at 128 Kbit/s (although a succession of high-quality 'stills' can be). Current kiosks tend to store their images locally, on CD-ROM or hard disk. A sensible design practice is to separate out the information and seduction components. The latter are stored on the disks, as stated, but the *information* (prices and availability, for instance) can be downloaded over the communications network. This avoids having to recreate the glossy catalogue each time a price change or discontinuation arises.

There is an obvious application of video-on-demand or high-speed cable modem technology for providing connection to seductive services at higher bandwidths. It is probably even possible to define the necessary standard of transmission quality required to provide a seductive service: it must approximate to that of the TV standards of the day. If we reflect on the lessons learned from paper catalogues, we see that there is no absolute standard for seductive quality; it is a moving target, and TV will probably be closest to the centre at any one time. Not surprisingly, some people consider that the future of home shopping lies with TV. For instance, industry analyst Dresdner Kleinwort Benson predict that sales from TV home shopping will top £1 billion per year in 2003, rising to £12 billion by 2009.

The mechanism for this will be the arrival of 'digital TV'. We can define 'digital' in a number of ways; it does not mean that the basic picture on the screen will undergo a significant change, but the signal arriving at the set will. It will be a digital datastream running at 2 Mbit/s or higher, thus at the very least preserving today's picture quality. It may originate from terrestrial or satellite broadcasts of digital signals, created by broadcast companies, from video-on-demand or cable modem entertainment or shopping channels, some of which may be identical to the off-air broadcasts, or from individual retailing groups that have Internet shopping pages.

Note that there is probably an even larger variety of programming options: for example, shops will create their own direct advertising and sales as well as distributing the activity via 'traditional' programme providers. Viewing may also become more personalised to the individual. To give an example of one of the interesting new opportunities: several years ago, the author and colleagues demonstrated the possibility of a 'stop me and buy one' service linked to video-on-demand. Because each household accesses a separate viewing of a programme, they can then pause it without affecting any other viewers. The idea of 'stop me and buy one' is that you can halt the action at any time to purchase items that you have seen on the screen. The (rather fanciful) example we demonstrated was buying the tie worn by an actor.

The ability to provide personalised channels in this way blurs the distinction between providing information/entertainment and advertising/selling. This is particularly true because of the high cost of creating and producing high-quality content. Much of the exciting audiovisual content will be reused: you may want to look at Australia from an information or interest point of view; alternatively you may want to buy a holiday there. In both cases the travelogue material will be the same surrounded by an appropriate shell, intended to inform you or to sell to you.

Although the arrival of the electronic shop can pose problems for retailers, in that the cost of entry to new competitors is drastically reduced, it is unlikely that home shopping will ever completely remove the need, or the desire, to go shopping: even today's shopping malls are places of entertainment and excitement rather than merely a way of buying essentials. The Mall in Edmonton, Canada, even has a dolphinarium, submarine rides, an ice rink, bungee jumping and just about everything else you can name.

They all still have check-out queues, though. All retailers want to remove queues, their surveys having shown them that every minute spent queuing equates approximately to one minute less choosing to buy. But providing extra check-outs costs money; so does providing large attractive areas in which to lay out the goods. Grocery retailing is capital intensive. From 1986 to 1991, the ratio of capital spending to pre-tax profit was in excess of 1.6 times. Most of this was in purchasing property. So, they are looking to networked solutions. In the future, either by a process of simply editing a list of staple items of food or by the use of an intelligent agent linked to electronic cupboards, a message will have been sent to the supermarket telling them what a customer requires.

They may deliver for a surcharge. European and US studies show that people, especially in families where both partners are employed, claim they are prepared to pay an extra 5% for home delivery. Alternatively, shops will pack the goods for collection on arrival. Although they occur a cost penalty in having to select and pack the goods, they may be able to offset this against a reduction in premium display space. Moreover, it will also help them in their stockholding: although they believe there is little they can do to improve the ability of their computer systems to predict the up-to-the-minute requirements for stock, if they can persuade customers to order the bulk of their goods in advance, this effectively gives the store's order handling processes additional time to make the forecast more accurate and to allow them to communicate the restocking requirements direct to the supplier and to the distribution network.

ENTERTAINMENT

Where physical goods are involved, there is still no technical magic that solves the fulfilment problem of delivering the product. However, there are a number of purchases which do not inherently require any material transfer. Of these, one of the biggest markets is in entertainment services. In many cases today, physical transportation is indeed involved: films are delivered to cinemas (and before that, film is sent from a number of studios to the studio HQ to be cut and spliced), videotape footage is still flown in to newsrooms, and the music industry sells a huge number of tapes and CDs. Once communication bandwidths of tens, or perhaps a few hundred, megabits per second become affordable, the need to transport video physically will disappear. Already, special-effects studios in Britain are transmitting their material over intercontinental links to Hollywood, cutting out turnround times and allowing for collaborative editing. Experiments are underway with electronic distribution to cinemas. We have already mentioned video-on-demand. This technique relies upon real-time video servers, typically with one computer located in each large metropolitan area and connected to a high-speed cable or telecoms network. The computers use inexpensive hard disk technology to store 1000 or so of the current favourite movies and TV programmes. Users access them through their TV sets, which are connected to the network. Because each user has separate access, the system behaves essentially like a video-cassette recorder, with pause and instant fast forward and back controls. The local server is backed up by a national service, which periodically downloads this week's favourites and can be used to supply other programmes that are not in the top 1000. As explained earlier, the service has an interactive backward channel, with the ability to make purchases or to take part in multi-user games.

Perhaps a cheaper alternative will be interactive digital TV, which cannot provide the fully on-demand services but will be run as 'stagger-cast', with programmes being repeated over a number of channels with, say, 20 minutes extra delay on each. Interaction from the viewer will be by telephony or cable connection.

As with teleshopping, the on-demand services create the possibility of individual, personalised programme selection. The user's personal preferences can easily be recorded. Targeted marketing becomes much more focused, down to a market size of one. The creators of films and TV programming will categorise them, by subject and nature, so that someone who enjoyed one film will be offered another of the same type. In due course, it will be possible to create automatic collages of similar material; suppose a news story breaks: perhaps artificial intelligence could be used to create an information magazine around it, giving an analysis of the circumstances in the place it occurred, at a depth that was appropriate to the individual viewer. It is important not to be too sophisticated: there is a danger that too much personalisation can give rise to a progressive narrowing of the user's interests and opinions. One could end up with a newspaper looking something like Figure 9.6. There will be the need to create an element of surprise, controversy, serendipity, and so on, which may mean human editorial intervention.

'Home-made Entertainment'

It is also possible that we shall see a fragmentation of the entertainment/information market, because of social factors combining with the low cost of publishing on the Internet. There are indications that social fragmentation is increasingly occurring, not as it once did around geographical proximity, because of difficulties of travel, but

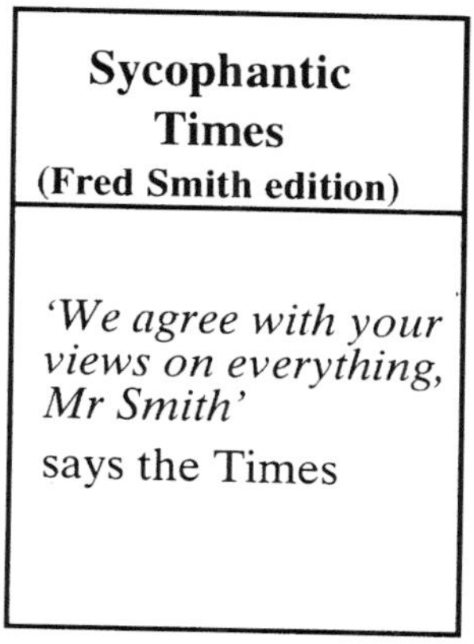

> **Sycophantic Times**
> **(Fred Smith edition)**
>
> 'We agree with your views on everything, Mr Smith'
>
> says the Times

Figure 9.6 The 'ultimate' personal newspaper

rather around common interest, irrespective of where one is. Combined with this are attempts to create common cultures on a much less than global scale, as happens, for example, in the way that popular music is fragmenting into different, differentiated types, rather than the homogeneous crazes of 20 years ago. It is now possible for communities of interest to create 'fanzines' (magazines for specific groups of fans) extremely easily and cheaply, in the form of a Web page. If a source of funding to meet the production costs is required, much of it may come from advertising rather than from user subscriptions. If we substitute 'proximity of interest' in place of geographical proximity, we can see that there is a resemblance to the free magazines advertising local product and services that have flooded the market in recent years. Their quality is now determined more by the talent and commitment of their creators than by the other costs of production. It is also important to realise that content creation is not simply one-way, from the 'official' authors out to recipients. Instead, e-mails and discussion groups generated by the latter can often swamp the 'official' material. The interactive nature of the Web adds a new dimension to publishing. Information reuse and packaging may be the key to success in the future.

Games

The continuing fall in prices of wideband circuits that we can expect from the introduction of new technology will soon provide opportunities for new sporting activities. Already many greyhound racing tracks in London are connected together via satellite links, so that racegoers can all be brought together, virtually, to watch (and bet on) races at other venues. It is cost alone which prohibits this from being more widely applied. We could even imagine virtual contests, where indoor athletic meetings take place at a number of stadia at the same time.

The USA video games market in 1993 amounted to $6.5 billion and continues to grow, with an estimated 70 million games machines in use. Most games are played on isolated machines, but trials with networked games, where several people compete at a distance, have proved very popular, apart from the cost of telephone connection charges which are widely perceived as off-putting. The continuing reduction in these charges, and the possibility of moving towards a charging regime which is partly subscription-based and not wholly determined by connect time, are likely to increase the take-up quite significantly.

It is worth mentioning that the bit-rate required, even for games with the highest quality multimedia, is not necessarily very significant. Consider the case of a Wild West 'shoot-out' game, where the two contestants move through a highly realistic, virtual town, seeking out each other for showdown. It does not really matter how graphically the images are rendered, not even if they are three-dimensional; the scenes are generated in the local terminals, that is, local to each player. All that is transmitted over the network are the geographical coordinates of the other player. This is sufficient to allow local generation of the other party's 'persona' (the character they play in the game). Thus, rates of a few hundred bits per second are all that is required. Of course, we can make the game more interesting – and potentially more lucrative to the network provider – by sending not just the coordinates but also the distant player's real face at data rates of a few tens or hundreds of bits each second (approximately a standard voice-grade call). This face would be electronically melded with the virtual reality body of the gunman, so you would have the satisfaction of shooting at someone you know.

DO PUBS HAVE A FUTURE?

One of the attractive features of the pub (or church or social club) is the feeling of belonging. In order to belong, you must be understood by people who should know something of your likes and dislikes.

That is one reason why these organisations have survived so successfully down the years and are likely to continue to do so. The 'local' is vanishing, but people are seeking out themed pubs which provide them with a range of entertainments to suit their taste. At the same time publicans are trying to cut costs, fight off the problem of crime and diversify into a market not solely reliant upon alcohol but yet sufficiently different from dance halls and traditional restaurants.

One way they can do this is to trade on the loyalty issue. An electronic means of achieving this is with some form of smartcard. Perhaps this can be plugged into the table you sit down at: it tells the bar staff who you are and what you like to drink. It may remind them it is about time they gave you your 21st birthday 'one on the house'. When you go on holiday, it and you will be accepted at all outlets in the chain. And it can be used as a credit or debit card to avoid the need for carrying large amounts of money and change. You may well have charged it up with prepayment cash from your home phone – or you may give one to your children, set-up to a fixed cash limit.

Of course there will also be pub games: these will be much more elaborate in scope than the ones we have at present. Perhaps they will use virtual reality and allow inter-pub competitions in a realistic environment, perhaps even against people in other countries. Many of the games that will eventually migrate to our homes, played on smaller screens and with cheaper equipment may well have started off in the pub of the future which can be an early adopter of wideband circuits and sophisticated computing power.

Pubs have increasingly become places for general celebration, including the provision of meals and private function rooms. The author and colleagues once demonstrated the 'virtual Christmas dinner', which allows people to sit down to celebrate Christmas at a table with a 3D screen that would allow full-size, realistic, two-way conversations with relatives or friends anywhere in the world.

Virtual Communities

Although the arrival of intelligent, affordable communications reduces the effects of geographical separation, the general concepts of 'distance' and proximity still have some meaning, at least in a figurative sense: we still expect organisations and services to be grouped together in some way, if their functions are related. Take an example from the World Wide Web: we are more likely to visit Web pages if they are directly linked to the page attracting our current attention. This is the reason why advertisers pay to have links to their sites placed on popular, frequently visited pages. A moment's thought will confirm that spatial metaphors are still strong, in cyber*space*: we 'navigate through', 'jump to', 'enter' and so on. At least two reasons can be given for the retention of the metaphor: firstly, though perhaps a transient phenomenon, the Internet is so slow that we do not want to waste time pursuing many links on the offchance of finding something useful; secondly, the spatial model represents the internal orderings our minds carry out, placing together in some form of spatial metaphor those things that are similar in concept.

Shopping Malls

Possibly one of the simplest examples is the 'shopping mall', a virtual retail park where a number of electronic shops are gathered together, in the sense that there is a top-level access site (on the Web or on a shopping channel of a video-on-demand or satellite TV service). The access site may display a 'map' of the mall, indicating which shops are available, and perhaps a set of common services such as some way of finding individual items, or a 'special offers' section that offers a set of combined purchases at a discounted rate. Apart from the online shops available, there is also the possibility of providing general advertising space or access to useful information – recipes, road traffic conditions, and the like. The claimed advantage of the mall concept (claimed at least by the creator of the mall, who will charge the individual retailers a fee) is that shoppers prefer to have a single point of contact that guarantees they will be able to find things that they know they want.

This might be described as part of an attempt to create a 'strong brand'; it is interesting to reflect on whether it is the individual brands on the site, or the brand of the mall, that are strong. In the retail community, 'brand ownership' is a vital commodity and there will be intense competition between mall owners and retailers in the mall, for dominance of brand. The mall owners may provide a transaction processing capability and, perhaps, electronic or even physical fulfilment. This may be done under their own brand or the retailer's (e.g. shop discount card) or by franchising the task on to a bank, a fulfilment company, etc. Because trading is online it also becomes easier for the mall owner to connect the customer directly to a manufacturer. This potential 'disintermediation' of the retailers is an issue that they are keenly aware of. (An early example can be seen in the concern of travel agents regarding direct sales by tour operators. Electronic trading makes this all the more feasible.)

VIRTUAL SCIENCE PARKS

The idea of providing collective services in a single mall is not restricted to retailing. One concept gaining ground is that of a 'virtual science park', frequently associated with a university or other location with a high technology background. The metaphor for this, which is sometimes actually instantiated as a virtual reality screen, is as shown in Figure 9.7.

Enterprises like this can demonstrate a strong brand-image of modernity and ability to harness new technology by going online to sell such services, as well as implying that their commercial operations gain synergy from the R&D strengths of the research organisation behind them. At the same time, the university, for example, is able to demonstrate its links with industry, an increasingly necessary feature.

In technology terms, what is required is a relatively fast connection between the partners on the science park, the critical element being the need to support their choice of interactive communications – videophone, video-conferencing, voice or simple e-mail. Given that they can be located anywhere, in principle, they probably benefit by sharing logical proximity to each other. (For example, they develop software for each other or require a particular type of research back-up from the university.)

Administration is a classic distributed IT management function,

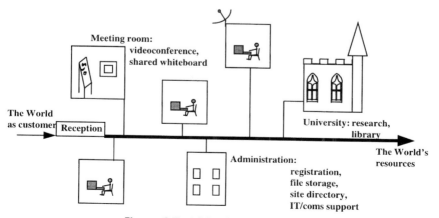

Figure 9.7 Virtual science park

involving the provision of network services and software support, together with the ability to set up (and take down, in the event of companies quitting the park) such services as *document reading rooms*, for the companies' clients to access, *video-conferencing*, billing, etc., and the element that binds together the whole activity – a *site directory*. This directory provides anyone who accesses the park with an easy way to navigate through the list of members and services on the site, and logs the names, addresses, functions and any other relevant details of the companies and employees in the park. One useful aspect of the directory in a classic university science park environment is in providing access to details of individual researchers, for example their published papers, the projects they have worked on, a current CV, etc. Notice that we distinguish between a directory and a database of such information. Many, if not all, the details will exist elsewhere, on other online servers. There is no need for the science park to copy this information into its own database; indeed this would lead to a maintenance problem, in that the original database and the science park version could get out of synchronisation. Instead, the science park directory merely holds pointers (the URLs) to the original database.

In practice, a fine balance has to be adopted between providing directory pointers and, instead, creating one's own database. Pointers say where something is, they do not say what it means in detail. For example, imagine providing access to research papers, consultancy reports or indeed any kind of document produced by members of the park. These will usually have titles, summaries, keywords, etc., but the way they are laid out may well make it impossible for a computer to extract that information automatically. How do we provide a 'mediation service' that allows us to search the data pointed to by the directory of the science park? There are a number of solutions, ranging from retyping all the data into a science park database (expensive) through asking the authors to cooperate by creating abstracts for filing into the park database or by meta-tagging (see Chapter 6) their existing documents (relying on their willingness to cooperate and their technical ability in the case of tagging), to simply letting an automatic search and indexing engine rampage through the existing data with erratic results.

VIRTUAL BUSINESS

The terms 'virtual business', 'virtual organisations' and 'virtual enterprise' are variously used to describe new ways of conducting business, based sometimes on new organisational structures and always involving computing and network technology.

There is a considerable market today in advice offered by expert consultants, and temporary business re-engineering teams instruct business executives in the received wisdom for improving the performance of their businesses by the application of computerisation and communications. Some of this advice is undoubtedly good; some of it (for example, the expectation that firms would increase the number of their suppliers through electronic trading) has been downright wrong. However, if one takes safety in numbers, there are a certain set of consistent views about business trends, which seem to have genuine relevance to technology:

- *Globalisation*: Buying and selling is becoming a worldwide activity. Even perishable goods such as fresh produce are fetched from three times as far as they were less than a decade ago. Small, 'local' butchers are on the Internet, selling prepacked specialities, worldwide. On a different scale, telecommunications companies are now major players in markets overseas from their origins.
- *Time to market*: Product life-cycles are shortening; over the last five years, US companies planned an increase in the rate of new product introduction of 21%.
- *Customisation*: Products will be designed in such a way that they can be adapted to the needs of individuals, at affordable cost to customer and supplier.
- *Customer loyalty*: Sales are no longer a single transaction; there is a need to retain customer loyalty by anticipating their needs and by affinity marketing.
- *Organisational flexibility*: Greater use of contracted workers to meet varying demand and skill or wage flexibility.
- *Creation of value networks*: Transient relationships and virtual enterprises that will exist for the duration of the project, and no longer.
- *Quality is a 'given'*: Quality will no longer be a differentiator; customers will flee from companies that do not deliver it.

Technology is seen as a facilitator in meeting these goals. In particular, intelligent communications is a way of marshalling the necessary organisational units together 'virtually', wherever they are geographically, within a support environment for controlling information and processes and promoting human interaction. A commonly used

Table 9.1 Technology in 'time' and 'place'

	Co-located (same place)	Distant (different place)
Synchronous (same time)	Meeting rooms Shared whiteboards and screens	Video-conference/videophone Cooperative design tools Group authoring
Asynchronous (different time)	File sharing Design tools Team rooms	Structured workflow E-mail, voicemail Bulletin boards

segmentation of the technology, in terms of the two dimensions of 'place' and 'time', is shown in Table 9.1.

It is also useful to consider 'virtual business' under a number of headings which, in general, are differentiated by the interaction of the individuals concerned and the nature of the trading relationships of their member organisations:

- *Teleworking*: We use this term to describe individuals who regularly work away from a central office (if one exists), on tasks which involve a degree of communication with others or with centralised data.
- *Cooperative working*: There are a number of activities, usually project-based, which involve marketing, design, development and production/maintenance teams, interacting to achieve a common goal. They may or may not be in the same organisation, but it is presumed that there is a high degree of prior commercial common purpose.
- *Virtual enterprises*: For our purposes, we define a virtual enterprise as one that comprises a number of units belonging to separate trading entities, that come together to achieve a specific goal, usually time-limited, that does not necessarily prohibit them competing in other areas.
- *Electronic commerce*: This has become essentially a generic term for purchasing of goods and services (either by end-customers, or through inter-business supply relationships), via the Internet. We have discussed it in some degree in the earlier section on retailing, and we shall cover other aspects in the sections on cooperative working and virtual working.

TELEWORKING

'Teleworking' was a fashionable word a few years ago, and we could be forgiven for concluding that it was nothing more than fashion, because it is seldom referred to today. However, we would be wrong; although it did not take off with the big bang that was predicted during the expansive 1980s when rising inner-city prices were expected to be the driver, it gradually, but steadily, has become a way of life for more and more people. There have always been jobs which involved working away from a fixed base – sales personnel, barristers, chartered accountants, as well as a number of people who work from their own homes, for example outworkers in the clothing industry. However, for the first time, through the use of computing and communication technology, it has begun to be possible to arrange for 'information workers' to operate away from a central office, usually in their own homes. Typically, these workers include computer programmers, 'call-centre' employees, direct marketing staff, indeed anyone with a data-processing and/or contact role. What has not yet happened is the wholesale replacement of the traditional office, as a centre for the work, by the completely stand-alone home workshop.

The reasons for this are predominately social; if people were machines, all we would require is a distributed operating environment and reliable communications. However, everyday experience confirms the results of formal studies that a lot more is required. There are a large number of activities that go together to create an efficient and tolerable office environment: we must closely interact with colleagues, supervisors and subordinates; we need face-to-face briefings on current issues, discussions of performance, training, collective problem solving, plus a number of informal activities (Figure 9.8).

Progress is being made with technological solutions to meet these requirements: electronic notice-boards and e-mail can now be

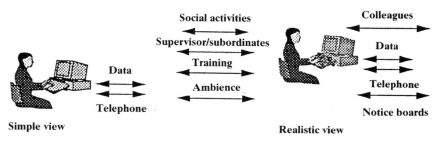

Figure 9.8 Simple and realistic needs of teleworkers

provided easily, as part of an integrated package. 'Groupware' products, for handling the creation and distribution of electronic forms, mean that business processes can be arranged to include the home-working members of a team. Additional applications can be written that allow the teleworker to go 'offline', in case of an emergency or a comfort break, without disruptions to any customer handling applications they might be involved in. In the case of these customer-handling tasks, the usual manner is for a central point to receive incoming calls and distribute them out to the teleworkers, using automatic call diversion, which balances the workloads and automatically finds a free line to a worker.

In many cases, it turns out that two telephone lines are required, one for the data being processed, and one for interactive communication, for example, with the supervisor. This makes basic-rate ISDN, with its two separate 64 Kbit/s channels, an attractive proposition, the more so because a single channel can support videophone. Video telephony may be more than an expensive luxury for teleworking communities: in one experiment, teleworkers were asked each week over a lengthy period to rate their satisfaction with their job. As shown in Figure 9.9, at one point they experienced a sharp dip in satisfaction.

This coincided with problems in providing the videophone service. It was noticed in this and other studies that video telephony comes

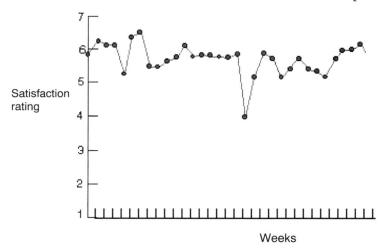

Figure 9.9 Response to the question 'How satisfied have you been with teleworking this week?': 7 = very satisfied, 6 = satisfied, 5 = fairly satisfied, 4 = neither satisfied nor dissatisfied, . . ., 1 = very dissatisfied. Reproduced by permission of British Telecommunications plc.

into its own when emotional contact is required; for example, in one case, rumours of closure of a branch resulted in a significant increase in the need for employees to 'look the supervisor in the eyes', even though this was done over a video link. Surprisingly, there does not appear to have been much work carried out into the technical requirements necessary to provide an adequate multimedia environment for full emotional support.

The technical barriers that remain are not unexpected; it is still quite expensive to provide basic connectivity between the business and its teleworkers. For computer programmers, the 64 or even 128 Kbit/s data-rate of ISDN does not compare favourably with the several megabits per second available for a LAN. The implementation of high-speed digital subscriber loop technology, either over copper telecoms cables or using cable modems, should contribute significantly in alleviating the cost and speed issues. There is also a maintenance problem with today's computers and their software: typically they require one support person to every 50–100 users; it is much more expensive to provide system support to outworkers than it is to people within the same building. Perhaps here is a good case for deploying network PCs (see Chapter 5). They can be simple, reliable machines, whose software is downloaded from a central server rather than being left to the teleworker to maintain. The availability of 2 Mbit/s links, even in one direction, from the centre to the teleworker would help significantly with system support and the downloading of files.

Most of the problems with teleworking are not to do with technology; they are the quite normal, human and organisational problems that beset any change to working practices. They will be resolved in due course and teleworking will undoubtedly grow in volume and scope.

COMPUTER-SUPPORTED COOPERATIVE WORKING (CSCW)

Although this heading gives the usual expansion of the acronym CSCW, it would be just as appropriate to render it as *communications-supported* cooperative working, because in all practical cases the people involved participate over a distance. CSCW is concerned with providing a supportive environment to distributed teams. This support is provided in a number of ways (Figure 9.10).

These three aspects are interrelated: processes (for example

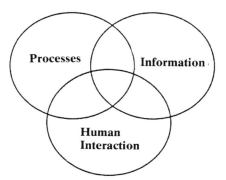

Figure 9.10 CSCW aspects

completing a purchase order) need to work on information and the information needs to be labelled in a way such that it can be understood by the process. People are nearly always key to businesses, whether virtual or not, and they must be supplied by the processes with the right information in the right form; they must also be able to collaborate on its interpretation in a manner which builds empathy and trust.

Technology allows interaction between humans to continue even over a distance. Well-established forms of communication such as the telephone and e-mail are obviously included. Increasingly, video telephony of some form or other is also used, to create a feeling of empathy within the teams. So far, however, its use is limited, partly to do with the lack of standardisation on the equipment available, but also because of cost or bit-rate problems. Using basic-rate ISDN, one can achieve reasonable quality images of individual people, filling a fraction of a PC screen, but the cost for this is that of a standard telephone call (or two calls if both of the ISDN 64 Kbit/s channels are used). At today's prices, this cost appears to be considered unacceptable, if one were to leave the call open all the time. Therefore, if the system is to be used as part of a collaborative working activity, one must go through the conscious action of setting up and clearing down the call every time one wishes to communicate with the distant party. Clearly, this is not conducive to seamless team-building. A further problem arises if multiple sites beyond two are involved: it is possible to join the video signals from a number of parties together in a multi-party bridge, but there are two unpleasant consequences; firstly, because delivery of the signal is limited to the fixed bandwidth of 128

Kbits, there can be considerable degradation of the quality; secondly, there is also the possibility of the introduction of significant further end-to-end delay.

An alternative method of delivery, at the present time, is via TCP/IP or similar computer protocol. This works well over a lightly loaded LAN or on a LAN where each workstation is separately wired to a central switch. In theory, systems of this type can adjust their bit rate requirements to suit the rate of change of information from picture frame to frame, and expand to cope with multisite use. Unfortunately this is no longer so, when we introduce long-distance travel over the Internet. The lack of quality of service guarantees usually means that poor quality images with bad sound synchronism result. (It is worthy of note that some studies of CSCW have tended to suggest that good quality voice is more important than quality images.) As with teleworking, high-speed digital delivery over the local cable may be beneficial, but, bearing in mind that the greatest benefit in the virtual team concept is the ability to operate over national or even intercontinental distances, it is clear that virtual businesses would benefit significantly from the promised low-cost, wideband, low end-to-end delay, international circuits we have been describing in earlier chapters. Given this provision of high quality, highly interactive multimedia, perhaps on wall-size screens, there is the possibility of conducting regular meetings with such a high level of empathy between the distant parties that real face-to-face meetings will be reduced.

ENGINEERING DESIGN AND MANUFACTURE

The technical areas of design, manufacturing, installation and maintenance have been among the earliest adopters of computer-aided support. In its early stages, it was single-user only, with little integration between the different stages and with processing done locally. Drawing offices adopted drawing packages that produced better results than hand-drawn layouts; design offices carried out simulations. Today all aspects of the process are beginning to come under control, in a consistent and integrated way, but there is still further development required in order to provide a fully satisfactory infrastructure for collaborative engineering involving dispersed, multi-disciplinary teams.

Documentation

Engineering documentation is not as boring a subject as the name suggests. In fact, some very exciting developments in process integration and maintenance are emerging, that are centred on the documentation.

Manufacturing industry has a long tradition of collaborative working across a number of sites; the construction of component parts in a number of factories and their bringing together in assembly shops elsewhere has been practised for centuries. So is the tradition that design and construction need not be colocated. In the past, this was achieved by tight document and drawing control: engineering drawing-offices were not just the places where the artwork was prepared, they also maintained the files of drawings, assembly instructions and processes. They were always in close contact with procurement functions in the organisation to ensure conformance of purchased components to the engineering specifications. Documentation is required to support a number of processes: production assembly, of course, but also for the control data to drive machine tools, the test patterns to drive equipment testing rigs, creation of maintenance schedules and manuals, stockholding, marketing and training information.

Increasingly, the introduction of computing led to the automation of the drawing process and to the filing of the documentation. In fact, the structure of documents, their flow and filing is a good representation of the collaborative processes that they support. Online information 'spaces' have been constructed on the basis of the way the documents are used. For instance, a virtual office, such as that in Figure 9.11, can be constructed automatically so that people within it are grouped together on the basis that they share frequent readership of the same document.

Members of the team can move around this virtual workspace, using a mouse or a joystick, having easy access to files of relevance to them, or to communication tools that let them set up audio or video calls to people in the virtual office.

Collaborative authoring tools are now in common use, so that many people can contribute individual views to one document; what has not yet emerged to any great extent is the single document that presents a selection of its many facets to individuals on the basis of their need. But this is coming: one example, created by the Sun computer company in California, is a technical manual that includes information on their operating system (Solaris) which exists in two versions, one for their Sparc workstation and one for an x86 machine.

Figure 9.11 A virtual office (reproduced by permission of BT Labs)

Users of the manual can click on a preference button, which will then let them view only the material relevant to either a Sparc or an x86.

This is one of the several advantages that can be gained from the use of document 'mark-up' and hierarchy techniques. In Chapter 6 we mentioned meta-tagging of documents (in the context of library categorisation) using the capabilities of Hypertext Mark-up Language (HTML), today's formatting language for Web pages. Meta-tags are instructions invisible to viewers of the document, but conveying formatting instructions to the computer's browser, e.g. that data was in a table, or headings should be represented as a certain height and colour. HTML is a user-friendly, cut-down version of SGML (Standardised Mark-up Language), which has been around for quite a while, and has been used particularly with US Department of Defense equipment documentation. SGML has proved too complex for everyday use, but a midway solution, XML (Extensible Mark-up Language), promises to strike a sensible balance between flexibility and ease of use. Whereas HTML principally makes use of a very restricted and predetermined set of meta-tags, XML allows users to define their own tags and use them to give meaning to their data. Thus, for example, the writer of a document could specify parts of it as relevant to 'maintenance' or 'production', and another user, who shared the common language of the tagging, could easily compose another document that included only those extracts from the first that were tagged as 'maintenance' information. Indeed, the other writer need not be a human; intelligent agents could be equipped with knowledge of the tags and, unaided, could put together the necessary documentation required for each stage in the production chain.

This is leading to the conversion of the data within the document into information with 'meaning' outside the document; a meaning that can be used to drive processes. A longer-term vision for this is a system which can combine:

- formally defined information, such as the new requirements 'the unit needs to carry a bigger payload'
- a set of assembly diagrams, whose rules are rigidly written down: 'part A fits on part B', etc.
- simulation results: 'the system runs too slowly'
- informal design decisions: 'I changed to a bigger motor, to solve the speed problem'

into one file, automatically linked and put under change control.

Another member of the design team, perhaps investigating why the tyres wore out, needs to see why the heavier motor was used. She can easily find the design decision and the original reason (a new requirement was specified) and can decide whether the requirement still holds, or whether to query the design decision (calling on the associated test data), or to redesign the tyres. This would be possible even for data distributed over heterogeneous databases, because the information types ('design' data, 'test' data, etc.) have all been specified using a consistent set of ideas.

This concept can be carried even further: abstract spatial concepts (such as 'no two items can occupy the same space') and laws of physics ('the heavier the load, the greater the friction on the wheels') are introduced.

This is called the creation of an 'ontology': a list of things in the universe under consideration, together with the ways they interact and with rules of logic that can reason about the likely impact of changes to the properties or things within that ontology.

Again, because we have decided on a limited number of 'things' and described how they relate and interact, the system can reason that 'bigger motor' implies 'bigger forces' implies a possible 'maintenance' issue, and can alert the maintenance designer to a possible problem and where to find it. This is an area at the leading edge of research, but there are some limited, experimental schemes which are beginning to show promise.

Virtual Prototyping

The process of going from concept to final product specification is an iterative one, involving intensive interaction between different groups of specialists, each of whom looks towards different target agendas bounded by different constraints (Figure 9.12).

Designers are first and foremost concerned with the aesthetics and appropriateness of the product as a concept; they may model in clay or draw freehand diagrams, or, increasingly, use computer graphics. A complementary part of their task is to convert these concept designs into detailed, dimensioned, engineering drawings. Again this is now commonly computerised.

For most large or intricate products or those exposed to environmental factors, the second phase is that of *analysis*: how will the product

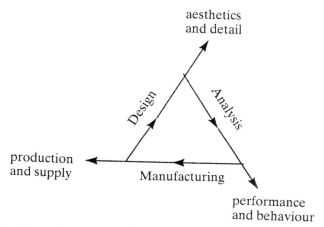

Figure 9.12 The product design process and the target agendas

behave in practice, can it withstand the likely stresses and strains imposed upon it, can it be maintained in service? This requires computational analysis, perhaps the solving of many large simultaneous equations, in the case of complex structures such as aircraft or bridges.

The final phase is concerned with the product's feasibility of *manufacture*. Is it designed in such a way that natural variations in its mass-produced components ('tolerances') will still allow it to fit together and function properly? These tolerances do not just relate to in-house manufacture and assembly; increasing specialisation means that few companies carry out all of this; more and more is contracted out to component and subassembly producers. Indeed, even the design and manufacturing functions may be split between separate organisations.

Given these varied viewpoints and agendas, it is amazing that successful product development happens at all. That it does is a tribute to the human ability to understand and negotiate between the conflicting constraints. However, much of this success has been due to intensive interaction within multidisciplinary teams, made possible by getting together in a single place, formally or informally, to bargain over the trade-offs to be gained in conflicting requirements, or simply to be able to understand the facts. To assist with this, expensive scale models are often used. If we are to achieve a comparable or, hopefully, an improved capability to achieve this on a global scale, we need ways of letting people meet 'virtually' and poke and prod virtual scale models, with the same facility that they do with the real thing. This is the quest of 'virtual prototyping', which relies on the interpersonal tools of video-conferencing and shared workspaces for its effectiveness; in particular, virtual prototyping is possible now because of improved visualisation techniques. Later in this chapter we discuss some requirements for such techniques, in the section on healthcare, where collaborative working is increasingly becoming the way forward, but the information and images manipulated within the domain of virtual engineering have some special characteristics. We are trying, in many cases, to remove the need for building real prototypes wherever possible. Real prototypes are expensive, often impossible to transport, and, not being to scale, sometimes not very accurate models of reality. For all these reasons, it would be preferable to create virtual equivalents.

When we do this, we have to realise that there is a great difference between the modelling tools currently available for sharing a view of what a product will look like (a design stylist's viewpoint) and the

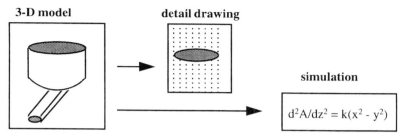

Figure 9.13 Current design-based engineering

tools which generate precise geometric and physical movement (the analysis and production/maintenance views). Existing modelling tools construct the images from polygonal shapes and then smooth them off and shade them. This is adequate for visualising the appearance of the product and makes it computationally feasible to display movement in real time, but it is not sufficient to drive cutting tools or to explore whether parts will fit together, particularly if one is looking at tolerances in machining, nor is it sufficient to study its aerodynamics or structural stiffness.

We particularly want to get away from the three-dimensional design acting as an uncontrolled basis for the other aspects of the prototype, the detailed engineering drawing and the structural analysis (Figure 9.13), into a situation where an underlying set of product information can be used by each of the process viewpoints – design, manufacturing, analysis – to create their own visualisations of the data and to introduce changes to it in a controlled manner (Figure 9.14).

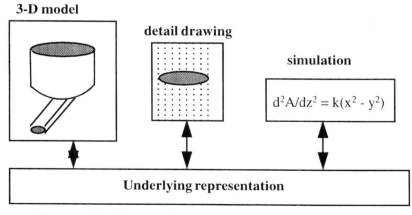

Figure 9.14 Integrated product development environment

The US CALS initiative is directed towards this idea of the introduction of electronic file exchange, integrated product databases and information models capable of including all information needed for design, manufacture and support of systems. Intended originally for the Department of Defense industries, it is now applied to other areas, such as automotive and aerospace, where long product lifetimes impose a considerable cost on maintenance procedures. CALS makes the product information accessible to authorised industry and contractors via electronic means. It can meet the needs of highly distributed organisations in which efficient communication and coordination of supply chains and third-party contractors have a significant effect on development and maintenance costs and delivery times.

CALS requires an effective mechanism for exchanging and sharing product data. Unfortunately the existence of many different engineering applications with non-standard file formats and information representations has made the exchange of product data difficult. To resolve these problems, the International Organisation for Standardisation (ISO) is currently in the process of developing an internationally agreed standard called the standard for the representation and exchange of product data (STEP) as IS0-10303.

Given a consistent way in which to store the data, we can build a collaborative prototyping system that will comprise services falling into a number of categories: sharing information, linking distributed people and programs, coordinating the team, integrating tools and services and capturing corporate history. The designers, analysts and production engineers can then call up the data to display it in the mode most suitable to their tasks.

Design

Basic three-dimensional design packages have been available for several years. They can be part of a pipeline display process, just as we describe in our healthcare examples elsewhere in this chapter, and in a collaborative engineering environment the viewed image can be constructed locally by each of the partners, from geometric information specified by one or more of them, or sent as a complete image file over the network. The former approach will greatly cut down the network data volume but requires local processing that is compatible across all the partners. If bit transport can be provided at a sufficiently low cost, then complete image transmission would offer increased flexibility in choice of collaborators.

Analysis

Analysis of complex products, for example their stress patterns or their dynamic behaviour under gravity and friction, is frequently carried out offline because of the computational load required, although, with the increasing power of new machines and the development of efficient mathematical methods, more and more we are seeing processing simulations being achieved in real-time. Where problems can be broken down into separable concurrent tasks, then distributed computer architectures such as CORBA (see Chapter 2) will make it easier for users to find a powerful processing resource without the need for expert computing skills.

Manufacture and Maintenance

These processes require accessibility, usability, and other ergonomic issues involving humans, to be taken into consideration.

Commercial tools exist that, for example, allow us to fit an articulated simulation of a human body into the modelling environment. We could imagine integrating our human figure into a set of designs from another design package that had modelled the three-dimensional view of a motor car. The figure could be placed within the car and manipulated to test for headroom, accessibility of controls, visibility and so on. It could even be submitted to dynamic testing, in a crash simulation, to measure the safety of the product.

In performing all these tests, there will always be a significant amount of design iteration; perhaps the most important part of this virtual design approach will be in the ability to present the proposed designs to other, geographically remote, co-workers, online, for their opinions and comments.

One recent example of such techniques was by the US Air Force who used the 'Jack' modelling system (based on research at the University of Pennsylvania) to model maintenance on the B1-B bomber in a demonstration of human factors analysis. Jack was used in tandem with the Human Operator Simulator (HOS) to produce animations of component change-out procedures, where it was found that human reach and the force required to change the component were factors requiring significant attention.

We might even imagine a scenario in the future, where, given the availability of better virtual reality tools such as data-gloves and virtual reality vision, we might simulate the physical difficulty experienced by the maintenance engineer when faced with locating the bolts and removing the oil filter, but current performance is generally not considered adequate enough for realistic use in an engineering environment.

Virtual Enterprises

In the case of cooperative design teams, we are entitled to assume that there is a reasonable likelihood that most members of the team share a set of common aims and are fully aware of the benefits of working to a common goal. Issues of retaining competitive advantage and hard-dealing over contracts and profit sharing are probably of less concern than that of getting things done, on time and within budget. Consequently, social purpose of the technological infrastructure is simply to maintain empathy, within an environment where trust has already been established, at least at the business level.

A more ambitious task is the creation of a virtual enterprise from a number of separate commercial entities who come together to achieve a common, but perhaps temporary, goal. They may even be competitors in other areas. How do they decide to come together and ensure that their own interests and that of the virtual enterprise will simultaneously be maximised? How can this be achieved in the case of geographically dispersed organisations? The answer lies in the way that technology can be used to link their processes and information together in a reliable and coherent way, and, in particular, how this technology can be used to enhance the trust between them.

Building Trust

A decade or so ago, many of the experts were predicting that the virtual businesses of the future would exhibit the behaviour of the marketplace: communication networks would make it much easier for the enterprise to pick and mix its members, and to discard them as soon as the current task was complete. The dominant partners would increasingly deal on a short-term basis of cheapest supply, with a wider and wider selection of suppliers. *This has not happened.* Why?

There are undoubtedly a number of reasons; one of them is that, quite simply, it is often not as easy as it might appear to connect together partners with disparate IT systems. But what also emerges from studies of this problem is the central role of 'trust', in the decision to select or reject a partner. Businesses, it appears, are behaving rationally in assessing their 'transaction costs' – the costs associated with dealing with other partners, and they are placing failure to get the lowest-price deal from their long-term partners as being much less important than the possibility of being let down by a new, lower-price, supplier.

Studies of trust confirm that evidence of past reliability is sought by companies looking for business partners. However, virtual businesses that are built up quickly to meet a temporary project and are then taken down, cannot easily supply evidence of past trustworthiness. Prospective partners must look to other factors that enhance trust. Organisations will be trustworthy if they need to be: it is difficult to cheat someone if you are forced to be open with your intentions. When we discussed retailing, we gave an example of this: suppliers having access to sales forecasts in return for revealing their stock-holding. An example from the banking industry involves a number of European banks which created a virtual company, based on a private, interbank network, whose sole task was to achieve rapid settlement of transactions from one currency to another. Transactions were placed in a common database and banks were committed there and then to settlement terms that could be instantly made available to customers by their indigenous banks, who had, in the past, to wait for several days to find out whether they themselves had struck a reasonable deal. The level of uncertainty between all parties was dramatically reduced, because the deals were solid and could not be repudiated.

Business theorists emphasise the need for organisations to share a clear vision of where they are and where they are going. Efforts to promote this are even more critical when the team has been recently put together to operate for a limited time on a specific task. There is a great deal of evidence to indicate that organisations and individuals that share a vision will trust each other. There is a need not just to read the memoranda and the mission statement, but also to 'look into the chief executive's eyes' and to hear the spoken words, explaining what is meant. The most effective method, a real walk-round the workplace, 'pressing the flesh' and meeting everyone, is often not practicable, but electronic substitutes can be used, 'Business TV', usually in the form of news bulletins and interviews with senior managers, is gaining in popularity with some organisations. The usual method of provision is via VSAT links, that sometimes are already in place for other tasks, such as sales data and pricing updates. Cheaper provision of 2 Mbit/s links and the development of software-only decoders will mean that business TV could be available at the desktop rather than on a big screen in the canteen. However, one-way broadcasts from the centre are not in themselves guaranteed to create trust among the workers; in fact, they can do just the opposite if the broadcasting is crass and untruthful. The broadcasters ought to be credible and appear competent and it would help if there were a backward channel, perhaps a voice link, that would allow the executive to be interrogated by a member of the audience. Senior executives should also be aware that e-mail and the telephone may well provide an alternative network to the official company line. Rather than object to this, they could realise that by empowering their workforce through changing the corporate magazine to an online, interactive service, they might gain a useful insight into company morale as well as new ideas for new products and services.

Business without People? – Intelligent Cooperative Agents

Within a company or enterprise, whether virtual or otherwise, we can identify a number of processes wherein two or more people have to work together to agree a solution. For example, there are contract negotiations between suppliers and users, there are work-scheduling tasks where a customer support centre needs to request resource from service engineers, there are rota planning tasks and so on. The people carrying out these jobs are increasingly supported by communications and computing tools, but they still make the final decisions themselves. Hovering around some of those tasks is the suspicion that they are not essentially very intellectually demanding and, indeed, that they might benefit from automation. However, the rigidity of standard packages ('workflow' is the term generally used) prohibits their use where the planning decisions need to be flexible: for instance, field force members can go sick or lack training in specific skills and these factors are currently best understood by human schedulers.

Challenging this viewpoint that human is best are more recent developments in the field of collaborative intelligent agents, mentioned in Chapter 2. These agents are modules of computer code that interact with their environment (principally comprising the other agents around them) through simple messages and, acting on the results of their actions, modify their future behaviour in order to achieve a specified goal. For instance, we could imagine an agent whose task is to negotiate a contract under the 'best' possible terms. We need to define what we mean by best: do we mean the lowest price for this deal, or the earliest delivery time? Are we trying to

APPLICATIONS

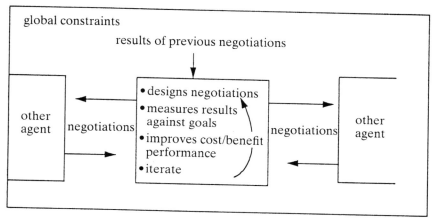

Figure 9.15 Adaptive, cooperative agents

optimise over a long-term relationship with the other party (in which case we are looking at average performance), or what? Schematically, the agent approach is given in Figure 9.15.

In the negotiation between agents we shall require a common language. If agent technologies become successful, we would expect standard dialogues within this language to emerge: contract negotiations, resource bargaining and so on. These dialogues would resemble 'quality of service agreements' that exist today between customers and suppliers of resources. For example, we would expect time limits for the supply of services to be agreed. Notice from the diagram that agents are supposed to improve their cost/benefit performance: agents are examples of a type of process known as an adaptive system. All adaptive systems require a way of computing their performance, in terms of maximising performance and minimising a cost function. It is in this way that they can carry out their tasks without explicit instructions.

Take the case of an agent that hires out a supply of support staff, e.g. to do telesales. The agent is trying to maximise the rate that it can charge for hiring out its staff, but is also aware that staff 'on the books' but not employed represent a cost. So, the agent will negotiate with other agents to hire out its staff at a certain price. If this results in too few being employed, it reduces the rate on the next iterations, until the rate for the job is arrived at. The process is one of Darwinian adaptation. It turns out that many problems involving simultaneous adjustment of many values, as would occur with a significant number of agent processes under negotiation, can automatically adjust themselves in this simple way, eventually to find a good solution, without having been given any detailed instruction beyond the cost/benefit optimisation rule.

It is rather similar to kicking a ball on a rough hillside. The height that we are at on the hill is a measure of the cost that we are trying to minimise. We kick the ball more or less at random and note whether it goes uphill or down. We base the direction of our next kick accordingly. Without us having to know the exact direction to kick the ball in order to get to the bottom of the hill, the ball eventually works its way down there.

Although the process is Darwinian, it need not be completely selfish. Seldom will an agent be working simply for itself; as the diagram shows, there are a set of global constraints surrounding groups of agents within which overall behaviour also needs to be optimised. For instance, a virtual enterprise might try to maximise the throughput of a collaborative manufacturing process. In this process, the only 'real' income is that arriving from external sales. It may be, therefore, that whilst agents within the enterprise negotiate with each other for resources, the measure of their success (and therefore the driving force behind their adaptation) will be their share of the profits that accrue to the whole enterprise, minus their own internal costs. The enterprise could decide to reward agents that turned out the maximum number of finished subassemblies for the minimum labour cost, or it could simply give everyone an equal share, or do something in between. The first option would encourage selfish, efficient behaviour, the second a more cooperative but perhaps less cost-conscious response. Quite complex and divergent behaviour could be expected to occur as the result of relatively simple alterations to the reward mechanisms.

Although the principle is simple, the construction of practical agent systems requires some thought and quite a significant amount of technology. Agents frequently have to negotiate with multiple simultaneous parties, thus the technology must support multitasking operations; in order to complete a single deal, it may occur that several subdeals must be set up ready to run, in which case programs must be capable of avoiding deadlocks (a process waiting for a process that is waiting for it) and so on. Also, if agents are to be expected to operate across separate organisations, there needs to be some forum in which they can advertise and specify their wares, and a technical environment wherein they can be effectively put to work. One possible environment is that of CORBA, as described in Chapter 2. CORBA is naturally designed for running a number of separate, but communicating, processes across distributed and heterogeneous computers. The CORBA Object Request Broker and the Trading Services offer a means whereby agents can represent the services they offer, in a consistent way.

HEALTHCARE AND DISABILITY

Collectively, we are growing old and for this and other reasons, we are spending more and more of our national budgets on medical health and care. The figures for the UK are summarised in graphical form in Figure 9.16.

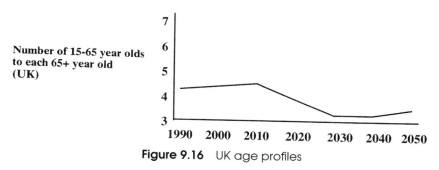

Figure 9.16 UK age profiles

The range of equipment available to doctors is growing year by year, as is the cost. Doctors themselves are expensive to train and specialisation means duplicating their skills across the world. Healthcare has always been an area of technological innovation and we are on the threshold of some major developments, brought about by computing and communications.

Remote Examination and Surgery

In our justification for the provision of broadband circuits we pointed out that providing access to specialist consultants was extremely expensive. This has already led to real live use of telecommunications to alleviate the problem. The range of medical images is extensive: X-ray plates, reconstructed images from magnetic resonance, axial tomography, PET scanners, ultrasound, pathology microscopy, laparoscopy, endoscopy and so on. So far, examples have mostly been for still images, which can be sent over basic-rate ISDN circuits at 128 Kbit/s, for examination at centres of expertise. Cautiously, experiments are underway with moving images too. Experiments have been held in which foetal scans have been viewed simultaneously in Queen Charlotte's Hospital in London and St Mary's Hospital in the Isle of Wight without loss of quality or definition. Basic-rate ISDN has also been used for obtaining real-time advice from consultants who remotely view images from endoscopy, and three-dimensional images of knee operations have been sent to other consultants for online comment.

The US Department of Defense has released details of actual cases involving the transmission of images, including the remote diagnosis of serious illness, successfully carried out using 'a very high resolution video system'. In another case, the death of a child was averted through the use of an electronic digital stethoscope to capture and relay his heart murmur from Italy to a specialist in Germany.

In some areas, the replacement of human surgeons by robots is producing beneficial results. The robots are not making intelligent decisions or carrying out fully autonomous actions – these still reside with the surgeon – but robot cutting devices can move more smoothly and with fewer and smaller tremors than scalpels held in human hands. The tools are guided to the site by the surgeon who, typically, views the scene on a large monitor and controls the operation by means of large handles whose positions are read by the system's computer. The pictures for the monitor are supplied by a miniature camera. In one experiment, control of the camera was achieved using recognition of simple voice commands. Tests of the use of this technique for coronary bypass have been carried out on pigs' hearts. Ultimately it may be possible to insert the camera and the surgical instruments into small incisions, perhaps less than a centimetre long, to carry out microsurgery with a precision and minimum intrusion well beyond that of unaided human operations.

So far, in experiments of this type, the surgeon and the patient have been collocated, but there is no reason in principle why this need always be so. The problem at the moment, apart from the natural reluctance to launch into a new and untried service, is the significant end-to-end delay imposed by the image coding techniques required to compress the moving video picture into an affordable bandwidth. This can be of the order of a few hundred milliseconds – far too much for real-time surgery. On the other hand, a broadband network that allowed a few megabits per second to be transmitted would permit a coder with much less latency (delay). Fixed line connections and satellite transmission are both possible solutions. If a very high speed circuit were available, say hundreds of megabits per second, the end-to-end delay would be less than the gap between the surgeon's movement and his recognition of its consequences. A further possibility would be to provide high quality three-dimensional imaging.

Surgical Applications of Collaborative Visualisation

It is also becoming increasingly the norm that surgery, just like any other highly skilled task, is becoming more and more a collaborative activity, carried out by interdisciplinary teams that may be geographically dispersed. A number of software systems have been developed to assist in collaborative working, including the very important task of visualisation, which is a means of presenting complex information in an easy to understand manner by representing its pictorially. The information may relate to real spatial parameters, such as X-ray photographs, or it may be simply a way of displaying a three-dimensional graph of results such as blood pressure, temperature and heart rate.

In the past, the controller and viewer of the data and images was a single individual, usually with locally held data and processing. As we have said, this is giving way to a collaborative approach. There are essentially two ways in which this can be carried out, depending on whether the system simply passes control and coordination information between the collaborators, or continually sends image data as well, while the visualisation is under development (Figure 9.17).

The control and coordination of the data is often achieved by providing a control token, which is passed from user to user, so that, at any one time, only one person is in charge of the processing of the data. An example of this is shown in Figure 9.18, where two doctors at different sites are simultaneously examining a view of the patient's skull.

In order to allow them to discuss features in the images it proves very useful for each to have a pointer. The figure shows one of them using a pointer to indicate an anomaly in the patient; the location of the pointer is held in a file of data common to both doctors so they can both see it.

Quite complex token-passing and coordination data passing methods have been developed. For instance, it is possible to ensure that actions carried out by one collaborator can be transmitted to the others, so that, provided each holds a copy of the data and the processing modules, they can run the pipelined operations on it locally, thus

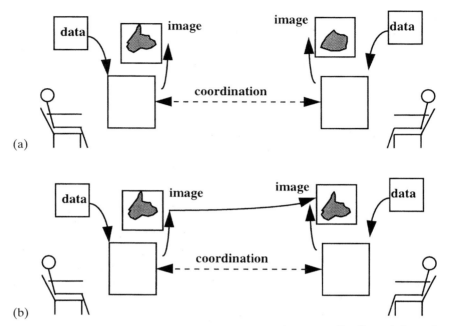

Figure 9.17 Options for visualisation: (a) passing coordination data only; (b) passing coordination data and image

avoiding the need to pass large volumes of completed image back and forth. Some systems hold an 'audit trial' of all the operations carried out on the data, so that they can be replayed, either by the originator or by the collaborators. This audit trail must be associated with a common data set. Integrated with a memo facility, it is a good way to retain historical records regarding the reason why decisions were made. With the increasingly litigious nature of healthcare, this may become more important.

The future direction in collaborative visualisation will be towards increasing diversity of the individual viewpoints and datasets, at the same time improving the coordination control and its user interface. The system should be flexible, allowing any number of participants to enter and leave the session, as would be the case in a healthcare multiple consultancy. Each participant will run his or her own manipulation on a dataset, which may be a superset, a subset, or completely disjoint from the date of other collaborators.

For example, one doctor has access to data showing structural information such as bones and soft tissue; another has data on functional

APPLICATIONS 345

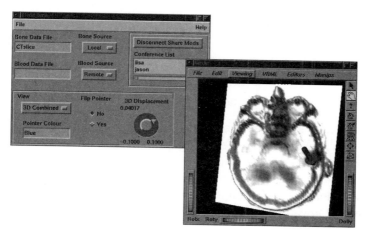

View of User 1

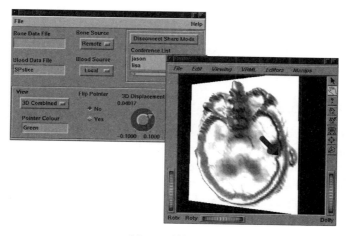

View of User 1

Figure 9.18 Sharing a common image set of medical data (reproduced by permission of Dr J. Wood)

information such as blood flow. They can combine this into a single graph relating blood flow to location within specific types of tissue. Ownership of the separate data on structure and flow is still retained by the corresponding specialist and cannot be modified by the other.

The data management process should allow data from any part of the pipeline (from raw data to completed image) to be transmitted and received by any partner. In the healthcare environment, hierarchies are important – often senior colleagues instruct juniors – and there needs to be a way of preserving this hierarchy in the rules for manipulating the data.

Transmission Considerations

Systems such as the one discussed above generate large quantities of data; X-ray, CT, etc. Files frequently contain tens of megabits of data, and compression techniques are dangerous as they might generate artefacts that cause undue concern or needless surgery, or conceal necessary detail. The data will be processed locally, as well as globally, and for cost and compatibility reasons will rely on standard computer protocols such as TCP/IP for its delivery, at least in the medium term. However, quality of service becomes a serious issue if the data is required in real time. The activity is one which will capitalise on the emergence of systems such as ATM, which will carry the data with low delay and in IP or any other reasonable format. Also, because the data transfer requirements are very 'bursty', packet systems such as ATM potentially offer cost advantages over permanent, fixed-bandwidth circuits.

'Full-sensory and Extra-sensory' Telemedicine

So far we have considered ways of remotely examining data, particularly visual, and the carrying out of remote physical manipulation. But there are other tasks carried out as part of diagnosis and general post-operative care that also might be achieved remotely, by providing telemetry of human and extra-human sensing. In Chapter 5 we mentioned artificial noses that could detect medical condition in patients and we also explained how the sensation of touch and the measurement of texture might be carried out remotely. There are other parameters which could be measured from a distance: heartbeat, blood pressure, the behaviour of the glandular systems, and so on. In the USA, patients discharged from hospital after heart attacks or surgery are sometimes wired up to monitoring equipment which stores and analyses their heart rate and rhythm. Although this is usually recorded locally, it can be transmitted over the wired or mobile network. It is even possible to operate in the reverse mode, in the event of heart failure: a cardiac resuscitator can be controlled remotely, acting on the basis of the transmitted data on the state of the victim's heartbeat. In general, none of these measures requires anything more than a narrowband channel – but it needs to be reliable, of course. There is also the need for a range of inexpensive and standardised sensors, to interface with the network equipment and the control system in the hospital. There is considerable potential for simultaneously improving the quality of life of patients and in reducing costs, through the use of telemetry, but natural caution and the need for standards are likely to be a restraining force on its rapid introduction.

Healthcare and Welfare in the Home

One of the aims of a successful healthcare service should be to keep people at home, rather than in hospital. Three aspects of this are (a) prophylactic diagnosis, intended to spot problems before they become serious, (b) post-operative support, and (c) sheltered support for the elderly and others, who would be in danger if left on their own.

Monitoring over the network is an obvious option. Already, within hospitals, it is possible for drug and medicine-dependent people, for example diabetic sufferers, to be monitored by remote computers that can administer and optimise dosage. There is every reason to believe that this technique will be extended to allow some of them to return home, and a simple dial-up facility provided to allow the hospital and the monitoring equipment to keep in touch. Other parameters such as heartbeat and blood pressure could also be monitored in this way. One of the issues is the development of inexpensive and standardised equipment for this purpose, since healthcare is a cost-critical area.

The Caring Home

Often people whose health is deteriorating change their patterns of behaviour before they are diagnosed as ill. This is particularly true of elderly people. One suggestion that has shown some promise when trialled experimentally is the idea of remote monitoring of behaviour to look for significant changes. Imagine a house with a number of its facilities connected to the network, via a 'home-bus' that could be hardwired or operated by radio. Things such as the individual room lights, the electric cooker, the washing machine, the bath taps, even the lavatory, could be connected. Their usage patterns could be logged and transmitted regularly to the healthcare centre, where they could be compared, manually or by artificial intelligence, with past history.

Significant variations could be flagged in order to inspire a visit to the occupant.

Of course, steps would have to be taken to avoid a 'big brother' impression of the service, but it could provide a significant increase in the opportunity for independent living for a large number of people.

THE VIRTUAL SOCIETY

In this chapter we have considered only a tiny sample of the applications that might be made possible by the arrival of intelligent, wideband communication networks. Even from these we can see that the potential impact on our lives could be immense. There is little doubt that any of the technology described above will be available soon. But that is not the end of the story: the question remains as to how quickly it will be taken up. There are formidable barriers to rapid progress towards a largely 'virtual' rather than geographically based society. Tasks that are 'important' are thereby also 'critical' and users will need to be convinced that technology solutions are consistently reliable. Surveys of owners of small businesses regularly report that they will not adopt new technology before they have seen it used effectively by others; how much stronger will be these acts of prudence in the case of telesurgery? In comparison with research into the technological issues, surprisingly little has been done into the social factors of distributed collaborative working. Perhaps it is not possible to do this, without it already becoming a fact. This makes it very difficult to predict the pace or direction of change. Hardly surprisingly, we can probably assume that the earlier adopters will be in the distribution of work that is currently considered non-critical and is already done by contract labour, probably in the areas of information management, where there is already a reasonably comfortable attitude to technology. Perhaps less obviously, medicine does also seem to be taking up the technology quite rapidly; this is probably because of the extreme scarcity and cost of expert resources.

As for transport telematics, this is an issue for governments and is surrounded by political issues. We need to remind ourselves that chaos theory teaches that many outcomes are fundamentally unpredictable!

10

Conclusions

At the beginning of this book, we presented claims and counter-claims regarding the reality of building an 'information society', based on advances in computing and communications. In the chapters that followed, we looked at developments in these technologies, attempting to assess the current state of the art, what was possible and what needed to be resolved. In this chapter we come to the hard part: how to present a concise 'auditor's report'.

Horrible visions confront auditors at such times, aware as they are of the ignominy of the scientist who proclaimed that atomic power was a myth, the Postmaster General who said that the telephone was 'unnecessary' and, in the case of the author of this book, his confident prediction in 1969 that self-service petrol stations would never take off ('reduced quality of service; too dangerous, will fail regulatory rules...').

Nevertheless, with the health warning that no one can be expected to get the future right without bags of luck, the analyses carried out in this book lead to some reasonable predictions.

WHAT'S REQUIRED

First of all, let us try to recap on what the market requires of the technology, in order to build an information society. Figure 10.1 attempts to segment the service requirements into a number of different facets. It shows that there are three distinct 'business' (that is, not 'technical')

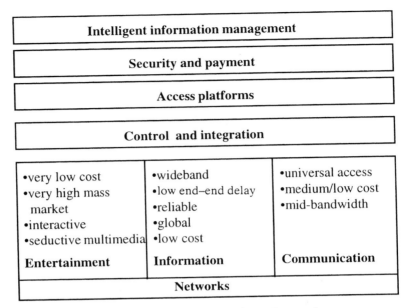

Figure 10.1 The layers of requirement

networks: one broadly concerned with mass entertainment, one concerned with interpersonal communication, and a third more generally concerned with carrying information of all types. These businesses will not completely converge, because they are selling services which are at least partially differentiated, although the technologies that deliver them may themselves have converged, in order to achieve benefits of scale and reuse.

- *The 'entertainment network'.* This feeds the domestic market and, as such, needs to deliver into the home. Here the battleground is between cable and digital broadcasting.
- *The 'communication network'.* Here we are referring to networks for interactive communication: basic telephony as well as videophony. This is the 'mobile' market and will be fought over by satellite and terrestrial radio, as well as traditional line-based telephony.
- *The 'information network'.* We have slightly extended our ICE definition of the information network (see Chapter 1) to cover the transport of any kind of data through the core or main network, that connects major switching/routing points.

Working inwards towards the core, we can set some timeframes and identify some technology directions.

ACCESS NETWORKS FOR 0–5 YEARS

The entertainment market, which mainly involves delivery to domestic premises, sets some severe constraints, in terms of (one-way) bit-rates and economy. If this market takes off, the compensation will be that huge volumes of equipment may create economies of scale, but so far this has been a belief held only by the satellite broadcasters and, perhaps, some cable companies. Telecoms companies have a dilemma: make do with Digital Subscriber Loop technology (Chapter 2), operating over existing pairs of wires, or go for an optical fibre solution, involving billions of pounds of investment for the UK. It is likely that, for the next five years or so, they will make do, although DSL technology involves some fairly complex terminating equipment, and it will be difficult to tackle effectively some of the electromagnetic interference generated by the system. Consequently, it could turn out to be an expensive stop-gap, necessary only as a way of retaining market share until a better solution arrives. Competing with them will be digital broadcasts of 2 Mbit/s and hybrid fibre/coaxial cable delivery from the cable TV companies. There may be some attempts to provide radio access at 2 Mbit/s from local, pole-top transmitters, provided by radio-access telecoms companies, but it is unlikely to pose a major threat in this timescale.

Ideally, the competing companies would like to use the same installed line-plant for 'information networks', particularly Internet services, at least to the home. Broadcasting would, at first sight, appear to be totally unusable for this: it is non-selective because of its broadcast nature, and does not possess a backward channel. However, it can be combined effectively with a telephony-grade line, and, provided only a selected subset of the entire Web is needed, can provide a multimedia equivalent of 'Ceefax/Oracle' Teletext. ISDN at 128 or 64 Kbit/s offers a mid-speed return channel, and it may be that ISDN can also provide a short-term solution for Internet access over the next few years. However, the multimedia capability at this data-rate is quite restricted and telecoms companies will face severe competition from the modems of cable companies, which should provide full-screen moving video at 2 Mbit/s.

Access for business will continue to move over to fibre as has been happening for several years. It is likely that, within the five-year timeframe, the transmission techniques will remain fairly 'conventional', but wavelength division multiplex may appear on what are effectively point-to-point circuits. ATM will become quite widely available, as a service, although a full range of ATM service qualities may not be publicly purchasable. ATM will not be 'at the desktop' within this period, except in pilot installations.

People at home will communicate with the world, as before, over voice telephones and via personal computers. ISDN at 64/128 Kbit/s may provide them with video-telephony, provided the market for it exists (this will require price reductions), or as part of their teleworking interaction. The radio telephone market will continue to expand, particularly for digital telephones and data terminals. Prices will fall, putting pressure on fixed line charges, because of increased competition in the terrestrial mobile market and, possibly, nudged at their top-end pricing, by satellite, particularly for very long-distance calls (as satellite methods are distance-insensitive), although this will happen towards the end of the five-year period.

ACCESS NETWORKS IN 5–10 YEARS

New entrants offering local loop delivery of entertainment and information via cable will be constructing a significant part of it using optical fibre. The conversion point from optical to electrical (i.e. from fibre to coaxial cable or copper wire pairs) will move closer and closer to the individual homes. There are two main reasons why the case for providing fibre direct into the home is, to date, rather weak: the cost, much of which is 'labour', of replacing the existing copper-based equipment, and the cost of the optical components (but not the fibre). If we are to get fibre to the home, then there will have to be a drop of at least one order of magnitude in the cost of the optical components. This is not impossible – indeed, it might be argued that the mass-market costs of DSL and fibre terminals are not dissimilar and fibre components are simpler and might eventually be cheaper – but going for fibre to the home will require commitment and faith in customer demand for wideband services. Nevertheless, we would expect to see some greenfield sites with fibre to the home before the end of the decade.

Radio and satellite systems will be trying to make significant inroads into the higher end of the business market for communications and for wideband information networks. There will also be attempts to provide local loop bypass, at 2 Mbit/s rates, using low-power, short-range transceivers, mounted on poles or buildings. Satellite will be a good solution for providing access in areas where there is no line system already installed, but it may run into national protectionism where it threatens to bypass the fixed network. A number of doubts have also been cast on the ability of satellite networks to cope with excess traffic volumes (e.g. in the case of national disasters such as a large earthquake) and on their ability to handle multicast transmissions. Although the potential returns could be great, satellite looks, at present, to be a relatively high-risk investment.

CORE NETWORKS IN 0–5 YEARS

Driven by their failure to deliver guaranteed quality of service at acceptable prices, computer and telecommunications networks have been forced to seek an accommodation. Computer networking is growing much faster than telephony, and data traffic will soon overtake voice in total bits per second. But, compared with telecommunications, computer networks are designed on a small scale, with local rather than global design rules, and they fail significantly in addressing standards and, especially, in quality of service. Their democratic, local rather than global, routing of traffic introduces lost packets and end-to-end delay that is unacceptable for tomorrow's multimedia services.

Telecommunications has responded with ATM – a rather unloved child that, however, has no real current rival. ATM does offer a compromise: no longer must you pay for a fixed, fast bit-rate circuit to carry your bursty traffic, which averages a much lower rate. Instead, you can have a range of quality-of-service offerings that adjust to the nature of your traffic (although no one has yet really worked out how or what to charge you). The good news is that ATM should guarantee this throughput rate and at an end-to-end delay which is fully acceptable.

The computer market will not wait for the traditionally cumbersome roll-out of telecoms services such as those entirely based on ATM. It is harnessed to Internet TCP/IP for its transmission protocols and to Ethernet for its LAN transmission method. Ethernet speeds are approaching the gigabit per second level and the tinkering with the Internet routers is beginning to provide the shadow of a quality-of-service definition. One probable event is the growth in number and scale (both geographical and in terms of size) of groups of private large-scale 'Intranet' networks. They will have only limited connectivity to the rest of the world but, within their own domain, they will be interconnected by whatever is cheapest and fastest, even across continents. A major uncertainty will be the relative balance between the provision of 'universal' connectivity through the traditional, centralist approach of telecommunications (which favours technology such as ATM) versus the islands of excellence approach, based on the interconnectivity of LANs.

As we have mentioned in earlier chapters, one major feature of computer networks that have their roots in LAN-outwards design is the lack of a sound quality-of-service definition. TCP was designed as a damage-limitation protocol, intended to support a minimum level of service for IP packets across the anarchy of the Internet. That is, the anarchic problem belongs to the networks, not to IP. Provided we could control the boundary of the wide area network, for example by using access control based on ATM, as described in Chapter 2, then there is a good case to be made for switching systems, within the wide area, to make use of IP routing methods rather than the switching principles of ATM. So, given the dynamism and growth in intelligent-terminal-based traffic compared with the staid progress of telephony, it is likely that even in the centralist, public networks full-blown ATM-based telecommunications will have to operate alongside IP-operating-over-ATM, if ATM is really to make headway. A solution will also be found to the problem of large-scale multicasting, for virtual business applications and for entertainment. Although no real competitor is visible today, it would not be too surprising if some unhistorically constrained equipment vendor were to break ranks and announce a competitor to ATM that was based firmly on the needs of the computer industry, rather than, as is true of ATM, something with its roots in voice telephony. It is still likely that this would be a packet-switched virtual circuit architecture, as is ATM. On the other hand, *if* ATM can get itself established within the next five years, it will be in a relatively strong position to be the *de facto* standard for at least the decade.

As far as the basic technology for carrying the traffic is concerned, the main network will, of course, be optical, and wavelength division multiplex in its 'dense' form (i.e. hundreds rather than 10 or so optical channels per fibre) will be used where there is a demand for extra capacity, but the switching/routing functions will still be electrical.

CORE NETWORKS IN 5–10 YEARS

Historically, protocols always proliferate in number and complexity as new applications rise up and bump their heads against the limitations of the technology. In the longer term, the arrival of limitless bandwidth will remove the heat from the debate. There will still have to be some set of conventions laid down so that terminals can address each other and understand how to interpret the data, but systems such as optical WDM would significantly simplify things. It may turn out that it is not the end-to-end 'infinite' bit-rate that fibre promises that is its real advantage: instead it may be that it simply provides enough 'slack' to allow us to get away with lightweight protocols and still achieve high quality of service and good interoperability. Nevertheless, the high capital investment involved may well mean that it will be significantly more than a decade before we see the core network becoming fully oriented towards an optical design, let alone being the dominant implementation, and legacy systems based on electronic/copper design principles will have to be integrated into the optical network for many years after that.

NETWORKS 10+ YEARS AHEAD

Optical switching/routing will mean that circuit and packet switching exchanges will be dramatically reduced in number, in principle five or six for the whole of the UK. Once the core network goes 'optical', in design strategy as well as implementation, there will significant advantages in network cost, maintenance and functionality, in extending the core performance out into the local loop, with fibre to the customer premises. Gigabit rates will be available at quite modest prices (reductions of at least one order of magnitude) if the market wants them. Technology and price may no longer be the deciding factors.

CONTROL AND INTEGRATION, 0–5 YEARS

Telecommunications control and the Internet will progressively become integrated. It will be possible well within five years to connect Internet name servers and intelligent network service control points. (So that, for example, a failed call attempt to a person who is not at home could automatically trigger an e-mail message, saying 'Couldn't get you; phone me!' from the caller to the called party's mailbox of the moment, and for the recipient of the message to set up a phone call back, to the correct number, pulled off the Internet name server.)

Towards the end of this period, networks will be designed (and some new ones implemented) using object-oriented architectures such as CORBA and TINA (see Chapter 2), and the distinction between computer network and telecoms network control will have become distinctly blurred. A number of business-critical processes will be running in distributed mode. Serious, 'live' experiments will be carried out, within this period, into the use of distributed control and agent technology, for network management.

CONTROL AND INTEGRATION, 5–10 YEARS

Efforts in standardisation will mean that, within the decade mobile and fixed-line telephony will be extensively integrated, sharing services and handsets (although the differing standards in Europe, the USA and the Far East may hold this back somewhat). It is perhaps doubtful that the distinction between connectionless and connection-oriented will have disappeared completely, as they have different virtues, but there will have emerged a common set of standards for control of 'computing' and 'telecommunications' networks and the latter distinction will have been lost.

It is likely that the experimental results in using distributed control by agents will be positive enough to lead to a modest amount being piloted within this period.

ACCESS PLATFORMS, 0–10 YEARS

Mobile units will acquire some multimedia capability within the decade (provided the market wants it), but it is likely that, whereas mobile and fixed telephony will converge, multimedia line services will drift away from line-based telephony towards computing and consumer products (although including mobile 'digital assistants'): see Figure 10.2.

With the start of the millennium, the domestic market will see a highly public battle for domination as the 'home platform' for multimedia services, between the computer and the TV set, although this may be more a marketing ploy than for real. The issue will be determined by user preference: will the information richness of the World Wide Web supplant the professionalism of interactive digital television? The chances are that manufacturing improvements will

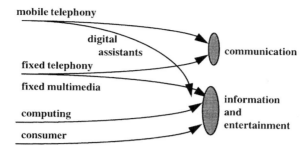

Figure 10.2 Different convergences

reduce prices and increase functionality, so that there will be room enough for two products (at least): a ubiquitous home computer, perhaps with a simplified interface, and a digital interactive TV set.

The consumer market is intensely price-sensitive, because of discretionary purchasing and the ever-present possibility of product substitution – the holiday for the computer, a camcorder or a new TV. Given this fact, there may be a market opportunity for a minimalist computer – the PC-lite, or network computer – built to a price and capable of downloading its software as required over a 2 Mbit/s link.

Terminal technology will see a serious rival to the cathode ray tube in the first five years of the twenty-first century. Screens will increase markedly in size, for the same price. Three-dimensional television will remain a gimmick for many years to come, but a significant minority of business applications (e.g. in manufacturing or medicine) will begin to make serious use of it within five years.

Some novel interfaces will be developed for special applications and, within five years and resulting from developments in artificial intelligence, terminals and networks will be able to interpret simple gestures and will respond via 'avatars' – artificial personalities – tailored to users' preferences.

SECURITY AND PAYMENT

It must be understood that security and payment mechanisms will never be good enough – they never have been! However, within the next five years, there will be stable methods for ensuring a high degree of security. There may well be a jittery period where developments of some of today's research in finding weaknesses in public-key systems are at least partially successful, but confidence will be restored. There will be a number of serious pilot, live deployments of biometric devices, within five years, of which retina scanning is likely to be one of the leading technologies. Smartcard technology, standards and applications will come of age within five years, and online electronic payment will account for a few percent of purchases. Some applications, such as road-tolling, may adopt electronic payment as their principal mechanism. Again within five years, governments will have to consider very seriously the issues of electronic money and international electronic purchasing.

INTELLIGENT MANAGEMENT OF INFORMATION

This is at once perhaps the most intriguing and most difficult area in which to predict progress. In the 'less intelligent' (i.e. that requiring the least application of *artificial* intelligence) activities, we shall see significant developments, well within five years, in the standardisation of labelling/marking-up/tagging of information, to make it more suitable for machine handling. This is essentially 'EDI for the Web' and will require human beings to develop the process models and agree the definitions. This will be supported by automated authoring tools that, whilst not employing much that could be described as 'deep' reasoning, will certainly make it much easier to manage information creation and to communicate between applications.

Machine 'interpretation' of language ('understanding' is too deep a word) will make steady rather than spectacular gains, over at least the next five years, and knowledgeable pessimists would say much longer than that. Machines will be able to identify the domain of interest of a document and tailor their dictionaries accordingly, but for the foreseeable future (greater than 10 years) they will frequently make serious mistakes in a non-human manner. This will apply all the stronger to artificial translation, which, however, will gradually be used for previewing and for a significant number of non-critical tasks. Speech recognition will follow behind these developments, at a rather slower pace. The 'easy' problem (see Chapter 6) of speech synthesis will steadily improve, so that, within a decade, the quality will be sufficient to generate a defamation suit if someone chooses to synthesise another person's voice, saying something they never said.

Image recognition will progress along a similar steady path: gesture recognition, with hands constrained within a reasonably behaved background, will soon become a usable but not 100% accurate technique for simple control of a system. Facial recognition will give adequate initial screening within populations of hundreds or even thousands, but not for critical verification purposes, and no theoretical breakthrough can be spotted on the horizon as yet. Simple multimedia classification and mark-up, as described in Chapter 6, will be progressively available within five years if a market for it exists.

One big problem with describing progress in the above area is that they are all very application dependent. It is actually very hard to decide whether a specific task is indeed 'difficult' or 'simple'. The author may be permitted an observation, based on some research he carried out some years ago in aspects of speech processing: it was recognised by everyone that, on the tasks in question, machines performed very badly; what it was difficult to make people realise was that people were also performing very badly. More generally, it turns out that a lot of what we do is very predictable and not at all clever, and there are tasks that computers are better than us at doing. The great and, the author believes, genuine hope for artificial intelligence solutions is not a belief that machine intelligence will become very high, but that surprisingly many 'intelligent' human activities are pretty dumb. With this consideration in mind, it is predicted that there will be a significant number of 'intelligent' tasks delegated to computers; for example, we would expect to see intelligent agents used seriously in service-level, and even contractual, negotiations between parts of a virtual organisation, across the network within the next five years. We have also made the rather lugubrious prediction that the current employment boom in call-centre jobs is seriously threatened by the introduction of online services that require no manual intervention. It is not difficult to imagine that casual employment which, at the bottom end of the scale, pays less than £5 per hour, requires little skill and could be replaced by an automaton.

FURTHER OUT YET

Even trying to predict likely developments within a 10-year span is quite difficult. To try to go beyond that period in any detail would be pointless. However, it might be interesting, at least, to do a small amount of technology spotting. We could ask, for instance, 'Which radical technology, related to communications, is likely to have the biggest impact on the world in 20 years' time?'. There are a number of possible contenders, including a major breakthrough in artificial intelligence, as discussed above, something radical in the bio-computing field, or communications-related activities brought about by research into how to avoid a celestial body colliding with the Earth. But in the author's opinion, it might be particularly worth taking out a small bet on 'nanotechnology' – the ability to make tiny machines, perhaps only a few thousands or even hundreds of molecules in size. Already millimetric-sized motors, cogs, levers, etc., have been manufactured experimentally. In principle, assemblies of such components could carry out complex tasks; for instance, tiny machines could be implanted in blood vessels and guided to sites in the body where they could carry out surgery. If it were possible to produce the machines cheaply enough, perhaps they could operate in 'gangs', cooperating by simple messages that made them behave like a single 'self-organising system', for example to transport heavy loads around a warehouse by 'rippling' in a controlled manner, like the feet of a millipede. In order to operate, there will be a need to provide communication both between the machines themselves and between them and central controllers.

In this regard, it is worth remembering that communications on its own has limitations. British intelligence was able to read the German cipher traffic even at an early stage of the war, but at that time all they had gained was an awareness of what was going to happen to them; they had no means to fight back.

These nano-machines would provide muscle alongside the artificial brains of future networks.

NO LONGER LIMITED BY TECHNOLOGY?

The overall conclusion of the audit is that very soon, say within 10–15 years, we shall have wide access to affordable networks of effectively infinite carrying capacity and computers with powers in excess of 1000 times those of today. Moreover, there are technologies on the horizon that suggest there may be additional mileage in Moore's Law even after that time. So, stretching ahead of us, in one generation, is the possibility that our communication and processing needs will not be limited by the technology.

However, there will be three very important limitations on the use to which we put this technology, and in the value we get from it: firstly, the technical problem outlined above and discussed in detail in Chapter 6 regarding the much slower progress in artificial intelligence; secondly, we do not yet have the experience to produce really effective products that substitute communications for physical transport – we do not yet know how to use multimedia conferencing to *replace*, rather than *supplement*, real meetings (or, indeed, understand whether it is possible). Finally, we do not understand the impact and implications that the new technology is going to have on the structure of our society – the extent to which new jobs will be created, and where, or the actions of governments, regulators, consumers, etc., in response to changed opportunities and threats.

In analysing power and intelligence in relation to information networks, we can see that we will soon have the power to do things we could not have dreamed of a decade ago, but the problem of limited intelligence, whether artificial or human, remains with us.

11

Appendix: Signal Theory

This appendix is intended to give just sufficient information to allow people unfamiliar with some of the fundamental principles of communication theory to understand such concepts as bandwidth, bitrate, sampling, etc., without requiring any extensive mathematical knowledge.

ANALOGUE AND DIGITAL SIGNALS

If we place a microphone in front of someone who is speaking, and look at the electrical signal that comes from it, we find we have a signal that looks something like Figure 11.1.

The vertical scale (in volts) is arbitrary and depends on the amount

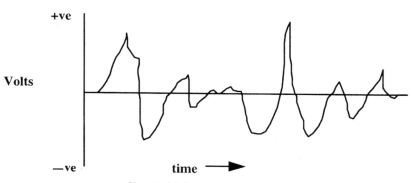

Figure 11.1 Analogue signal

of amplification in the system. The time period covered in the diagram is about 20 milliseconds. We see that the waveform is continuous: it goes from one place to the next without any jumps or other discontinuities. The waveform is a good copy of the changes above and below air pressure that occur near someone's mouth when they speak. That is, the signal is an *analogue* of the original source.

Now, for reasons that should become clearer later, we decide that we do not need to know the value of the signal at every instant in time. Instead, we represent the signal as a set of *time samples* (Figure 11.2).

Usually, as shown, we sample at regular intervals. Thereafter, we know the value of the signal only at the time samples; we know nothing about what happens between the samples. Therefore, it can be seen that we have to decide upon a sampling rate that is good enough (i.e. fast enough) for whatever purpose we have in mind; if it is too infrequent we have too many possibilities in the gaps. We shall return to this problem later.

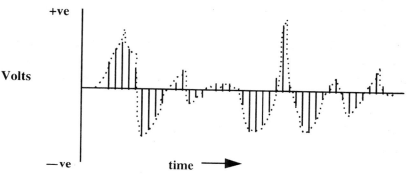

Figure 11.2 Time sampling

Now let us abuse our signal in a further way as shown in Figure 11.3. Not only are we going to break it up into time samples, we have also decided that its height (the value of the voltage at any one time) will be 'quantised' into a number of levels, to each of which we have assigned a binary number. So, in the figure, the first time sample is in level '000', the second in 001, third in 010, fourth in 110 and fifth in 101. Thus, we can represent our signal's values in time as a string of numbers: 000, 001, 010, 110, 101. That is, we have created a *digital signal*. We can see that each sample of the signal can be represented by three bits. Notice that we could find an infinite number of signals that all had the same binary values as our signal. We need to have a

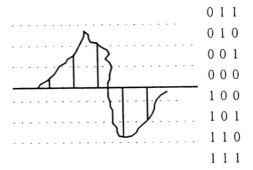

Figure 11.3 Quantised sample

sufficient number of quantisation levels in order to make sure that the other signals with the same binary values are sufficiently close to our signal that we can consider them to be identical for any purpose we want to put them to. For instance, if we consider telephone speech, sampled in the way described, we need to have an 8-bit quantisation in order to get sufficiently acceptable quality. (The number of levels equals 2 to the power of $8 = 256$.)

Note that the levels do not need to be equally spaced. In fact, for speech we have wider levels for highest positive and negative samples and levels closer together around zero. This tends to keep the error per sample a constant fraction of the sample size; this is subjectively the best solution.

So, if we have a signal which is quantised to N bits and we sample it in time at a rate of R samples per second, then:

$$\text{bit rate of signal} = N \times R \text{ bits per second}$$

We know that the quantisation, N, depends on how accurate we want our samples to be represented, and that depends on the necessary signal quality we want to present, but what about the choice of R? To answer that question, we need to tackle the problem from a slightly different angle.

THE CONCEPT OF FREQUENCY

Suppose we plot a graph of the vertical position of a weight suspended on a spring as it oscillates up and down after we pull it and let it go; let us also measure the sound wave from an organ pipe or from a gently plucked guitar string. In each case, the oscillation will eventually die

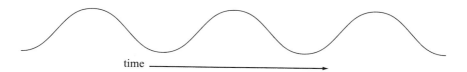

Figure 11.4 A 'smooth' oscillation

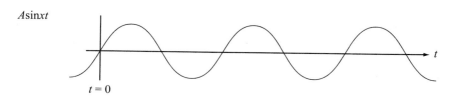

Figure 11.5 A sine wave

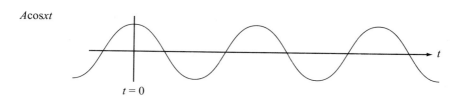

Figure 11.6 A cosine wave

down, but to a reasonable approximation it will have the form shown in Figure 11.4. It is a smooth shape, which we may remember from maths as being a sine wave (Figure 11.5) or, if we decide to start our clock at a different time, a cosine wave (Figure 11.6). (Sine and cosine have the same shape, but they are measured from different starting times.) We see that the shape repeats after a certain period of time. Let this time be T seconds. Then we define a quantity called the *frequency* of the sine/cosine wave, by the formula:

$$\text{frequency} = 1/T \text{ cycles/second or 'hertz'}$$

To put a size to it, the musical note often used by musicians to tune their instruments by, 'A above middle C', is standardised at 440 hertz (Hz).

Notice that frequency is defined only for sine and cosine waves. It turns out that it is not useful to associate a single frequency with waves of other shapes, such as our speech signal. But there is a way whereby we can associate a *range* of frequencies with waves of this type.

FREQUENCY ANALYSIS

Nearly 200 years ago, Jean-Baptiste Fourier, a French mathematician, proved that complex waves, such as our speech wave, can be shown to be made up of a sum of sine and cosine waves of different sizes and of frequencies f, $2f$, $3f$, $4f$ and so on. The lowest frequency, f, is known as the fundamental, and the multiplies of it are known as harmonics (Nf is the 'Nth harmonic'). So, if we represent this on a diagram, showing a pure sine tone as a single line (Figure 11.7), we can then represent a more complex signal as a set of frequencies of varying heights (Figure 11.8).

The range of frequencies, from lowest to highest, is known as the *bandwidth* of the signal. Alternatively, if we have a transmission channel (a coaxial cable, an optical fibre, a telecoms system with filtering on the input) which can pass only a certain range of frequencies, then this range is known as the *bandwidth of the channel*.

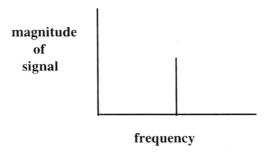

Figure 11.7 Frequency components of a pure tone

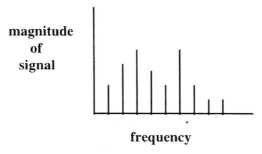

Figure 11.8 Frequency spectrum of a complex waveform

SAMPLING RATE

When we looked at time-sampling, we raised the question of the necessary value of the sampling rate, R. The reason why we need to have rates above a certain value is shown in Figure 11.9.

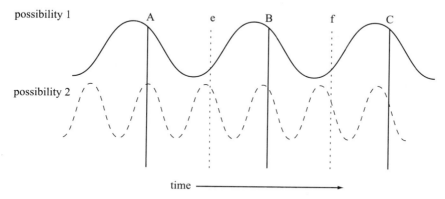

Figure 11.9 Ambiguity from slow sampling

Sampling the top signal at A, B, C, etc., is insufficiently fast to distinguish it from its second harmonic (shown below it, as a broken line). Obviously, if we are later to try to reconstruct the original signal from its time-sampled values, we would have no way of resolving this ambiguity. As the figure shows, we must sample faster, at A, e, B, f, C, in order to resolve the two. It can be shown (but not here) that, for any signal of bandwidth B (in hertz), the minimum sampling rate required, R, is given by:

$$R = 2B$$

If we do so, it turns out that not only do we capture all the information in the signal that we are sampling, but it is also easy, by passing it through a filter which produces a smooth continuous interpolation between the time samples, to convert it back to its original, continuous (non-time-sampled) form.

TRANSMISSION BIT-RATE FOR SPEECH

Using the above formulae, we look at some of the implications for the digital transmission of speech and video.

The quality of speech on normal telephone lines can be described as adequate rather than high fidelity. Although speech covers a frequency spectrum from zero to about 10 000 Hz ('10 kilohertz' or '10 kHz'), only signals below about 3.4 kHz are transmitted over the telephone. The original reason for this was that, when all telephony was analogue and not always amplified along its path, the line and the telephone instrument both attenuated the higher frequencies. In more recent times, because it makes the digital handling of signals easier, a low-pass filter is used deliberately to restrict the signal to a bandwidth below 3.4 kHz.

From the formula in the last paragraph, this would mean that we need to sample the incoming speech at a rate at least 2×3400 samples per second. In practice, a rate of 8000 samples per second is invariably used.

Now, we stated some paragraphs earlier that the amplitude variation of speech samples is such that they can be quantised into 256 levels, which requires eight bits to describe the levels uniquely.

To send this digitised speech information over a communication channel, we need to use one that can support a bit rate of 8 bits/sample $\times$ 8000 samples per second, that is, *64 Kbit/s*. This has been the 'traditional' (i.e. without complicated compression techniques) rate at which a single telephone channel is sent over digital networks and is why the figure 64 Kbit/s, and multiples thereof, recurs throughout this book.

There are ways of reducing this figure of 64 Kbit/s quite considerably, by using the fact that it is possible to estimate the next sample of speech by looking at previous samples. These are extensively used in mobile radio, to give fully acceptable speech quality at bit rates of 16 or even 8 Kbit/s, but a fuller discussion is beyond the scope of this appendix.

TRANSMISSION BIT-RATE FOR VIDEO

When we come to examine TV signals, we see we are dealing with a much faster rate of delivery of information. The bandwidth of broadcast TV signals is around 5 MHz, therefore a sampling rate of twice that, 10 million samples per second, is required. Without doing anything very clever, we would need to use an 8-bit quantiser (as for speech, but in this case the levels are usually chosen to be of equal sizes at all signal levels). This gives us the not inconsiderable data-rate of 8 bits × 10 million samples per second = 80 Mbit/s. This would mean that one video signal would require the equivalent of more than 1000 speech channels (= 80 Mbit/s divided by 64 Kbit/s), which could be rather expensive.

Fortunately, there are ways to improve this significantly. They rely on the fact that, within any one TV frame and between any successive frames, quite a lot of the data does not change rapidly with position or time. Seldom, for instance, do we get a picture that looks like that shown in Figure 11.10.

This is a very 'busy' picture and, if we take a section through it, horizontally at A–B or vertically at C–D, we get the square waveform as shown, which varies very rapidly with position (rather than with time, as in our speech waveform). Therefore, we can think of this picture as containing a pattern which tends to have a lot of high 'spatial' frequency components (Figure 11.11). This is a very unlikely pattern. In most cases, the distribution of spatial frequencies is biased towards the low-frequency end ('A' in Figure 11.12). Very infrequently does the picture have a flat spectrum of spatial frequencies (curve 'B').

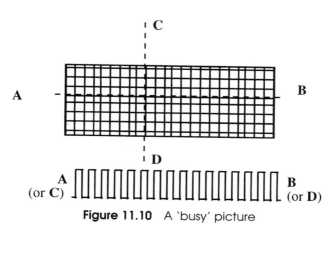

Figure 11.10 A 'busy' picture

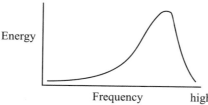

Figure 11.11 Frequency spectrum of 'busy' picture

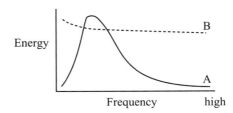

Figure 11.12 Frequency spectrum of 'typical' picture

A little thought may persuade us that not every part of the spatial frequency spectrum of any one image is equally important; the high-energy parts of it contribute more to the content of the image than the small parts. So, we can throw away the latter without losing much picture quality (Figure 11.13).

The picture (one timeframe of a TV image – there are about 25 per second) is stored in a buffer store. We apply an image processing algorithm to the data; typically, this will be a 'Discrete Cosine Transform', one way of computing the Fourier terms of the horizontal and vertical spatial spectra. Having computed the values, we impose a threshold, or simply select the top few values, then digitally code and transmit them. At the receive end, we process them through an inverse transformation to create the picture waveform. This may lack a few sharp edges, but by sending enough of the Fourier terms it usually bears a close resemblance to the original, without having to send so much data.

In fact, we usually do not carry out this operation on a single frame

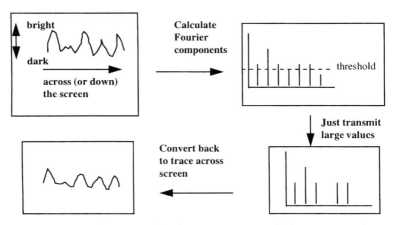

Figure 11.13 Removing low-energy spectral components

Frame 1   Frame 2   Difference

Figure 11.14 Interframe coding

of the picture. Just as we can take advantage of the fact that, within a frame, spatial changes are seldom as fast as they could be, we can also trade on the fact that the change from one frame to the next is also relatively slow (Figure 11.14).

Between frame 1 and frame 2 of Figure 11.14, the only part of the picture that has changed is the movement of the figure. (In fact, this movement has been exaggerated to make the point; it would seldom be as much as this between adjacent frames.) Now, if we have already sent frame 1, why send all of frame 2? Why not just send information on the difference between them and let the far end calculate frame 2 by adding this difference to frame 1? That is exactly what is done by several popular forms of image coding. They first compute the difference between successive frames and then code this difference, using, for example, the Discrete Cosine Transform, throwing away all but the largest values. These are then transmitted to the far end. Of course, in order to start things off, they have to send the first image in its entirety, and only thereafter send differences. This means that there will be a start-up phase, with consequent delay, while the larger volume of first-image data is being sent.

An additional amount of delay is also incurred because the rate of change of images, from frame to frame, is not constant. If the differences were sent immediately they were calculated, a variable bit-rate would be required. Instead, the data for several frames is held in a buffer and clocked out at a steady rate. Thereby, slow and fast changes of scene average out the required data rate. The delay introduced must not be greater than about 200 milliseconds, otherwise two-way video telephony begins to suffer. Clearly, systems for one-way transmission, such as for TV, are not limited to this size of delay.

These techniques of coding of signals, because they make use of the fact that information is essentially repeated (e.g. by not totally changing the brightness/colour picture from one point in the picture to the next, or from frame to frame), are known as 'redundancy coders'; subtracting previous values from current values is known as 'differential encoding', and since the data rate required to transmit the signal has been reduced, the coding is known as a 'compression' technique.

In this way, TV-quality images can be produced from data streams of less than 2 Mbit/s. That is, we can send high-quality moving pictures over the equivalent of 30 speech channels each carrying 64 Kbit/s.

CHANNEL BANDWIDTH

We have said above that we need to provide a channel that is wide enough to carry the width (in terms of frequency range) of signal that we want to transmit. We have also indicated that our signals will be digital signals. In many cases, they will not just be digital but also 'binary'. That is, as shown in Figure 11.15, we shall represent a digital word, say 101, by a string of pulses that can take either the value 1 or 0 (or 1 or −1).

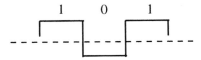

Figure 11.15 Perfect binary string

Now, each one of these bits is shown as being a perfect rectangle with infinitely sharp edges. If we were to carry out an analysis of one of these bits using Fourier's technique, we would find out that the series of frequencies that made a perfect recipe for the rectangle would go on to infinity. Now, no real transmission channel ever has an infinite bandwidth and we always want to work as close to its maximum as possible, in order to get the highest data rate. Let us try to operate at the lowest bandwidth possible (Figure 11.16).

In the figure, we see a bit-stream of alternate 1s and 0s, running at rate R. From Fourier's method, we can also see that the fundamental frequency of this pulse train is a sine wave of frequency $R/2$. (It can be shown that this is true also for more complex pulse trains.) So, we would think that all we need to have is a transmission channel that can pass frequencies up to $R/2$. However, we find that this is not true, when we look in detail at what happens when a single pulse passes through the channel (Figure 11.17).

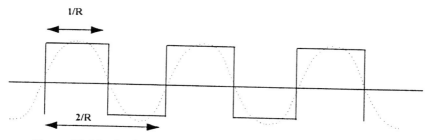

Figure 11.16 Perfect binary string and fundamental frequency

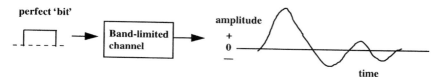

Figure 11.17 Pulse distortion in frequency-limited channel

A single pulse, of finite duration in time, when passed through a band-limited channel, emerges as a signal that oscillates back and forth above the zero line, dying away gradually. This is called 'ringing'. A pulse of the opposite polarity (a '0' rather than a '1') would, of course, first swing from zero to its maximum negative value. So, at first sight it would appear easy to distinguish one from the other, by simply measuring the value at the midpoint of the first swing. But what about the other bits in the data stream? They also generate ringing, and it is possible that the sum of all this ringing may drown out the effect of the current pulse we wish to measure. This ringing effect turns out to be particularly bad if the channel has a very sharp cut-off – that is, below a certain value it lets through all frequencies equally, but above a certain value (the 'cut-off frequency') it attenuates them effectively to zero, as in the continuous line of Figure 11.18).

It is better to choose a gentle roll-off, as shown by the dotted curve.

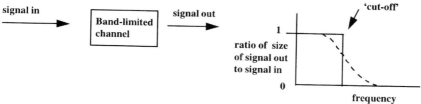

Figure 11.18 Bit-rate and channel bandwidth

This means that, for a given data rate, R, we require a bandwidth greater than $R/2$. In fact, we find that we need to add on a factor of around 0.77 to achieve reliable detection of the bits. That is why it is reasonable to remember that *the required channel bandwidth, in hertz, is approximately the bit-rate in bits per second.*

WHY DIGITAL?

We have said that the channel bandwidth needs to be roughly the same size as the bit-rate passing over it. Thus, for digitised speech at 64 Kbit/s we require a bandwidth of around 64 kHz. However, we also said that analogue speech can be contained within a bandwidth of about 3 kHz. So why bother to go digital at all? The answer lies in the fact that bandwidth is not the only parameter that describes the performance of a transmission channel. An equally important parameter is its 'dynamic range'. This is the space available between the maximum allowed size of signal, before the electronic circuits in the channel overload, and the background noise, below which the signal disappears, as shown in Figure 11.19.

These two limitations cause no problems if the channel is short. However, on long channels, such as a trunk telephone cable between two cities, there is a progressive attenuation along the path, such that we have to amplify the signal before it gets lost in the noise. Of course, any amplification we carry out amplifies not only the signal but the noise as well, and as we progress along the route, we add in more and more noise, eventually producing a signal which is very different from the one we started out with. But now consider a digital signal consisting of binary 1s and 0s. Unlike an analogue signal, at any one instant in time it can have only one of two values. It is not allowed to have anything in between. When we pass it down our transmission channel, it too gets attenuated and requires to be boosted in some way, to avoid it also disappearing in the noise. But we do not amplify

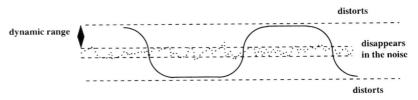

Figure 11.19 Limitations on signal size

it (and its accompanying noise). Instead, we make a decision as to whether it is meant to be a 1 or a 0 and, on the basis of our decision, we simply generate a perfect, new pulse, that travels onwards to the next regeneration point, and so on until it reaches the receiver. Provided we keep a certain minimum signal to noise ratio, we shall always be able to decide which of the two pulses to send, with vanishingly small chance of error. This is one of the advantages of digital transmission: it provides a much higher guarantee of received signal quality in situations where dynamic range is limited, the penalty being an increase in bandwidth.

The received belief of telecommunications network design over the past few decades has been that digital transmission is the correct way to operate. It has been claimed, however, that this viewpoint should be challenged with the arrival of optical networks, as it is easier to get some of the transmitters and receivers to operate in what is in some sense an analogue mode, and this, combined with the large dynamic range of fibre (particularly its low noise), may result in a move back to analogue transmission within some parts of the network. (It should be noted, however, that the data that are likely to be carried will mainly be computer data and therefore digital!) This is probably more than enough to say on this topic, which, however it excites some designers, is quite confusing for those with a casual interest.

12

Selected Further Reading

A comprehensive list that fully covered the topics in this book would be as daunting for the reader as for the author; instead I have selected a sample of books, articles and Web sites, generally suitable to non-specialists, from which it should be possible to follow up further references if required.

General Reading

A great deal of material at the appropriate level is to be found in the IEE's *Electronics and Communication Engineering Journal*. The *IEEE Communications Magazine* is a key source, approachable by semi-technical readers. (The *IEEE Proceedings* are more technical.) Similarly, for computing aspects, the special editions of the *ACM Computing Surveys* are highly approachable, as is the *Communications of the ACM* which often has special issues that are not too difficult.

2 New Network Architectures

Intelligent Networks

Dufour I (ed.), *Network Intelligence*, Chapman & Hall BT series, 1996. A comprehensive and realistic view of the opportunities and problems.

Sharp and Clegg, Advanced intelligent networks – now a reality, *Electronics and Communication Engineering Journal*, June 1994. A concise overview of technical and standards progress.

ATM

Handel R, Huber M and Schroder S, *ATM Networks*, Addison-Wesley, 1997. A detailed text.

Agents and Network Management

Appleby and Steward, *BT Tech. Journal*, vol 12, no 2.

Distributed Computing, CORBA and OMG

Siegel J, *CORBA – Fundamentals and Programming*, John Wiley & Sons, 1996.

Lewandski S M, Frameworks for component-based client/server computing, *ACM Computing Surveys*, March 1998. An excellent long paper, which deservedly won the ACM Surveys student prize, gives a clear (though perhaps subjective) presentation on the virtues of CORBA compared with other solutions.

The Object Management Group maintains a comprehensive Web page at www.omg.com

TINA

Lapierre *et al.*, TINA: a software architecture for telecommunication services, *Electrical Communication* (Alcatel), 3rd Quarter, 1994.

Rowbotham T and Yates M, Tina – a collaborative way forward, *British Telecommunications Engineering*, vol 15, part 1, April 1996.

IEE Conference Publication 451, *Sixth IEE Conference on Telecommunications*, Edinburgh, 1998. A good flavour can be got from a number of papers which apply TINA to issues such as mobile telephony and convergence.

3 Advanced Optical Fibre Networks

Fibre Systems

Geckler S, *Optical Fiber Transmission Systems*, Artech House, 1987.

Green P E Jr, *Fiber Optic Networks*, Prentice-Hall, 1993. An excellent in-depth text by a fibre aficionado.

Passive Optical Networks (PONs)

Cook A and Atern J, Optical fiber access – perspectives towards the 21st century, *IEEE Communications Magazine*, February 1994.

Mochida Y, Technologies for local-access fibering, *IEEE Communications Magazine*, February 1994.

Fibre Components

Arnold J M, Solitons in communications, *Electronics and Communication Engineering Journal*, April 1996.

Kashima N, *Passive Optical Components for Optical Fibre Transmission*, Artech House, 1995.

4 Radical Options for Radio

General Radio Issues

Stone W R (ed.), *Review of Radio Science 1990–1992*, Oxford University Press, 1993. Predominately technical explanations but approachable by non-specialists.

Cellular Telephony

IEEE Communications Magazine, April 1998. Devoted to the subject of locating mobile units in cellular networks.

Digital TV

Dobbie W, Digital video broadcasting standards, in Whyte B (ed.), *Multimedia Telecommunications*, BT series, Chapman & Hall, 1997.

Peltor H, Digital television broadcasting: the UK start-up, *IEE Review*, July 1998.

Satellites

Hadjitheodosiou M H, Coakley F P and Evans B G, Next generation multiservice VSAT networks, *Electronics and Communication Engineering Journal*, June 1997.

Kohn D M, Providing global broadband Internet access using low-earth-orbit satellites, *Computer Networks and ISDN Systems*, November 1997.

Roddy D, *Satellite Communications*, 2nd edn, McGraw-Hill, 1995.

The online pages of the various LEO partners contain detailed white papers and technical submissions to the regulators.

5 Accessing the Network

Social Aspects

Hirsch E and Silverstone R, *Consuming Technologies*, Routledge, 1992. Subtitled 'Media and information in domestic spaces', this is a good corrective to engineers who believe products are just about technology.

Mobile Terminals

Tuttlebee W, You'll never talk alone, *IEE Review*, May 1997.

Computer–Telephony Integration (CTI)

Swale R and Chesterman D, Computer–telephony integration, *BT Tech. Journal*, April 1995.

Chesterman and others have also given a simpler overview in *British Telecommunications Engineering*, July 1995.

Communications Week and *America's Networks* both carry frequent articles on CTI from an industry perspective.

Set-top Boxes

A book that I edited, *Multimedia Telecommunications*, BT series, Chapman & Hall, 1997, contains excellent chapters on broadcast services by Will Dobbie, and on set-top boxes by Bob Bissell and Adrian Eales

Hardware

Ezra D, Look, no glasses, *IEE Review*, September 1996. Survey of 3-D.

Gardner J W, Microsensations, *IEE Review*, September 1995. Includes the electronic nose.

Scholey N, Rechargeable batteries for mobile communications, *Electronics and Communication Engineering Journal*, June 1995.

Convergence

Brown L and Pattinson H, Information technology and telecommunications, *Management Decision*, vol 33, no 4, 1995.

There are also frequent articles in the technology sections of *The Economist*, *Financial Times* and *The Guardian*, and in the *New Scientist*, on set-top boxes and the convergence debate, covering not just the technical issues but also the critically important business, regulatory and international political constraints and concerns which technologists must take into account.

Network Computers

Again, this topic is as well covered in the general press as in specialist magazines. Many of the issues concern market-driven economics. For a polemical account which nevertheless contains compelling statistics, a book-in-progress, *CIOs are Dead Meat*, by Ron Cooke, promises to be worth reading.

Extracts are available, at the time of writing, on the World Wide Web at http://www.ncworldmag/com/new-10-1997/new-10-ciodm.html

Strategic Directions in Computing

ACM 50th anniversary issue – Strategic Directions in Computing Research, *ACM Computing Surveys*, vol 28, no 4, December 1996.

ATM Computer Architecture

McAuley D R and Leslie I M, The desk area network, *Electronics and Communication Engineering Journal*, June 1997.

6 Intelligent Management of Networked Information

ACM Transactions on Information Systems provides a consistent supply of good articles (technical, but approachable) on information management. Try, for example, the October 1997 edition which has two long papers on intelligent searching, with extensive references.

The Frame Problem in Artificial Intelligence

Pylyshyn Z (ed.), *The Robots Dilemma*, Ablex Publishing Corporation, New Jersey, 1988.

Speech and Language Processing

Keller E (ed.), *Fundamentals of Speech Synthesis and Speech Recognition*, John Wiley & Sons, 1994.
Wheedon C and Linggard R, *Speech and Language Processing*, Chapman & Hall, 1990.
A very good analysis of recognition systems, including hidden Markov modelling and practical results, is by professor Steve Young in *Institute of Acoustics Bulletin*, vol 20, no 5, 1995.

Neural Networks

Two good introductory texts are:
Dayhoff J E, *Neural Network Architectures*, International Thomson Computer Press, 1996.
Wasserman P D, *Neural Computing*, Van Nostrand Reinhold, 1989.

Image Processing and Recognition

Sonka M, Hlavac V and Boyle R D, *Image Processing, Analysis and Machine Vision*, Brooke Cole, Pacific Grove, CA, 1998. An undergraduate coursebook.

Morris R J and Hogg D C, Statistical models of object interaction, *IEEE Workshop*, Bombay, January 1998. Explains in detail the car-park crime example.

Intelligent Multiple Media Techniques

Smith M A and Kanade T, *Video Skimming for Quick Browsing Based on Audio and Image Characterisation*, Carnegie Mellon Computer Science, July 1995.

7 Technologies for a Wicked World

Anderson R J, Why cryptosystems fail, *Communications of the ACM*, November 1994. Still one of the best explanations of why things go wrong in practice.

Bhaskar K, *Computer Security*, NCC, Blackwell, 1993. Describes the ESPRIT-MARS project which looked at both the technical and operational aspects.

Cohen F B, *Protection and Security on the Information Superhighway*, John Wiley & Sons, 1997.

Fairhurst M C, Signature verification revisited, *Electronics and Communication Engineering Journal*, December 1997.

Koops B-J, Crypto regulation in Europe. Some key trends and issues, *Computer Networks and ISDN Systems*, November 1997. Good list of references to policy papers.

The Internet community is obsessed with security and human rights. Any public search-engine will provide more than enough sites. The European Commission has produced a number of thoughtful papers and lightly moderates discussion groups, at http://www.ispo.cec.be (in particular, see /eif/policy/97503.htm)

Smartcards

Dettmer R, Getting smarter, *IEE Review*, May 1998.

8 The Stakeholders

General

Beyond the Telephone, the Television and the PC, OFTEL, August 1995. This consultative document issued by the UK regulator is a thoughtful and thought-provoking introduction to the complex issues involved. With rather appealing modesty, it requests as many answers to its questions as it makes statements.

Green Paper on the Convergence of the Telecommunications, Media and Information Technology Sectors, and the Implications for Regulation towards an Information Society Approach, COM(97)623, European Commission, Brussels, 3 December 1997. Despite its wordy title, this is a readable paper on current EU official thinking.

Gerald W, *Telecommunication Policy for the Information Age*, Harvard University Press, 1994. People who suspect regulation to be a monolithically dull subject will find this book, by an academic secondee to the FCC, a surprisingly interesting and informative read.

Education on the Move

See references to a European research project, TOPILOT, in *Important Issues in Today's Telematics Research, Proceedings of the Telematics Application Programme*, Barcelona, 4–7 February 1998, published by DGXIII European Commission.

PICS

Miller J, Resnick P and Singer D, *Rating Services and Rating Systems (and their Machine Readable Descriptions)*, Internet Draft draft-pics-services-00.txt, 21 November 1995.

Copyright

Several papers in the *ICC Proceedings of the Multimedia Communications 93 Conference*, Banff, Alberta, April 1993, discuss copyright details in a non-specialist and interesting way.

Privacy and Interception

Several highly relevant papers in the *Proceedings of the 8th Joint European Networking Conference (JENC8)* are available from jenc8-sec@terena.nl

State Censorship

Peng Hwa Ang and Berlinda Nadarajan, Censorship and the Internet: a Singapore perspective. *Communications of the ACM*, June 1996.

9 Applications

Transport Telematics

Summers L, *The Strategic Deployment of Advanced Transport Telematics*, Centre for Exploitation of Science and Technology (CEST), October 1993.

Entertainment

Glover, R, Weber J and Melcher R A, The entertainment economy, *Business Week International*, 14 March 1994.

CSCW

Computer Networks and ISDN Systems, special edition, vol 29, no 15, November 1997, contains a number of useful papers selected from the 8th Joint European Networking Conference.

Wood J D, *Collaborative Visualisation*, PhD thesis, Department of Computer Science, University of Leeds, 1998. The example of collaborative medicine is taken from this useful thesis.

Virtual Enterprises

Hawryszkiewycz, I, *Designing the Networked Enterprise*, Artech House, 1997.

Electronic Commerce

Communications of the ACM, June 1996, contains a number of papers of direct relevance.

Hypertext Mark-up, SGML, HTML, XML

The best place to look for information and tutorials on XML is via the Web pages of the World Wide Web consortium, http://www.w3.org/MarkUp/

Collaborative Documentation and Ontological Reasoning

Gruber *et al.*, Towards a knowledge medium for collaborative product development, *Proceedings of the 2nd International Conference on Artificial Intelligence in Design*, Pittsburgh, June 1992, Kluwer Academic.

At the time of writing, there is also useful information on the Stanford Knowledge Systems Laboratory Web pages at http://www-ksl.stanford.edu

10 Conclusions

The technical and business press provides numerous articles predicting future trends. Peter Cochrane, Bonny Ralph and myself edited a series of non-specialist articles in each part of Volumes 13 and 14 of *British Telecommunications Engineering*, 1994 and 1995.

CONVAIR, a European project in the ACTS Programme. is looking at the evolution of the network: see http://www.cordis/lu/

Hurwicz M, Preparing for gigabit Ethernet, *Byte*, October 1997. Presents the latest in LAN technology.

Mace S, ATM's shrinking role, *Byte*, October 1997. Despite the title and the admitted computer bias, this is a very fair assessment of the medium-term future of networks

Futurology

Three views from the 'it's going to happen sooner than you think' school are:
Cochrane P, *Tips for Time Travellers*, Orion Business Books, London, 1997.
Gates B, *The Road Ahead*, Penguin, 1996.
Negreponte, N, *Being Digital*, Vintage Books, 1996.

11 Appendix: Signal Theory

If you have read Chapter 11 and are looking for further reading, the chances are that you are not a signal theory expert. So, rather than tackling one of the many modern texts, you may find it more helpful to look for things produced at a time when people had really begun to be able to explain the subject and before they became so knowledgeable that they made it too difficult. One of the most approachable and interesting books is *Symbols, Signals and Noise* by J R Pierce (Hutchinson, 1962). Treatment of signal processing forms only a part of it, but the rest is highly relevant to discussions of artificial intelligence. For a more formal introduction and feel for the subject, *Waveform Quantisation and Coding* edited by N S Jayant (IEEE Press, 1978), still takes a lot of beating.

INDEX

Note: *computing and telecommunications terms abound with abbreviations, which are more widely used than their expansions; consequently, this index uses abbreviations where they are the more commonly used form.*

0800 service 269
access
   conditional 279–280
   fair 263, 266, 268, 276
   universal 19
access control 225
accident 301
adaptive system 339
addressing 45–46
ADSL (Asymmetric Digital
   Subscriber Loop) 61–63, 81
advertising 282, 312
aerospace 334
agent, enquiry 150
agent, intelligent 25, 67–68, 70–72,
   305, 330, 338–341, 359
airline 301
airport 303
alliance 277
amplification 85, 87
amplification, optical 20
analogue 8, 33, 363
anchor point 218
animation 24
announcements, recorded 38
answering machine 11, 64, 123, 142
answering service 40
architecture 28–82
artificial intelligence 20, 23–7, 188,
   191, 209, 221, 357–361
   definition of 180
   for healthcare 347
   local 161
   meta-tagging by 289
artificial neural network 212–215, 255
artificial personality 216

AT+T 266
ATM (Asynchronous Transfer
   Mode) 13, 40, 52–60, 81, 87,
   345–6, 351–4
attack, nuclear 44
authentication 225, 243, 249, 254,
   266–7
automated order taking 205
automatic call diversion 324
automatic call generation 151
automotive 334
avatar 24, 357

B1–B bomber 335
bandwidth 32, 36, 49, 50, 362, 366
   channel 366, 372–4
   fixed 52
   optical 93, 112
   variable 52
banking 215, 337
barge people 288
base station 22, 120, 126, 131, 141–2
battery 153–155
beacon 304
behaviour, emergent 72
billing 16, 41, 62, 271, 276, 277, 283
bio-computing 360
biometrics 247, 249–257, 358
bit-rate 362
bit-rate, variable *see also* 'bandwidth,
   variable' 51
bit-rate-limited 59
bit-tax 298
blind people 289
block enciphering 236
Bogart, Humphrey 207

Bolognia 301
brand 318
bridge (in LAN) 42
broadcast-and-select 105, 106
broadcasting 279, 282
  digital 350, 351
  satellite (see also 'satellite') 15, 17
  terrestrial 15, 299, 311
browser 44, 68
BSM (Broadband Switched Mass-market Services) 272
BT (British Telecom) 264, 265, 271
bucket, leaky 56–57
bugging 242, 295
bus (electronic) 88
bus (motor) 301

cabinet, street 61, 87, 102
cable company see 'company, cable'
cable, coaxial 6, 8, 15, 31, 60, 61, 84–6, 94, 117, 166
cable, submarine 110, 118, 129, 228
call-centre 6, 287, 299, 323, 359
calling, cashless 38
CALS 334
capacity, channel 91
capacity, fibre 97
car 301
car park 216, 294, 303, 305
car radio 139
car-thief 216
cardiac resuscitator 346
CAT 83
catalogue (shopping) 144, 308–9
cathode ray tube 357
CD-ROM 272, 298, 311
CDMA 50, 132–133, 169
Ceefax 351
CELESTRI 136
cell (ATM) 53–54
cell (radio) 120, 122
cellular radio 120
censorship 291, 297
certification authority 247
CGI (Common Gateway Interface) 74
charges 6, 38
charging 269–270, 273–274, 279
charging, reverse 124

check-out 312
China 299
Chinese lottery 237–8
cinema 313
ciphertext 232, 233
circuit 49, 51
circuit, integrated 21, 22
circuit, switched 52
circus 288
citizen 262, 297, 298
classification 185, 187
classification, linear 250
CLI (Calling Line Identity) 117, 124, 142, 271
client 74, 76
client-server 41, 79
clockwork 155
close-up 220
code, downloadable 148
collaborative authoring 328
collaborative manufacturing 335, 340
collaborative visualisation 343–4
collaborative working 328, 348
commuting 5
company
  cable 9, 13, 60, 61, 62, 85, 127, 264, 271, 279, 350
  computer 13
  telecommunications 7, 9, 10, 11, 13, 19, 32, 37, 66, 126, 127, 263, 265, 268, 273, 274, 277, 351
competition 262, 275
compression
  data 84
  digital 128
  video 16
computer
  ATM 175–6
  games 138, 139, 140
  home 15, 66, 67, 75, 138, 139
  laptop 153, 155
  network(ed) 146–9, 165–6, 276, 277, 325
  personal 16–18, 21–3, 30, 50, 65, 143, 146, 153–4, 352
  portable 154
  virtual 23

INDEX

computing
   distributed 73–75, 80, 81
   parallel 238
   performance 21
   power 19
concentrator 31
confidentiality 224, 225
congestion, traffic 5
connection 36, 46–47, 51
connection-based 166, 175
connection-oriented 356
connectionism 213
connectionless 51, 166, 175, 356
consultant 223
content 273–274
contract 338
control, call 32, 35
control, congestion 45
convergence 18, 19, 26, 138, 275–6, 280, 285
convoy (vehicle) 303
cooperative design 327
copyright 26, 79, 281–4
CORBA (Common Object Request Broker Architecture) 25, 75, 335, 340–341, 356
CORBAfacilities 78
CORBAservices 78
cost 6, 30, 52, 56, 266–267, 269, 276
cost, transaction 337
couch potato 140
coupler, optical 92, 100
credit card 26
crime 26, 306
criminal 26, 217, 293
cross-border 275
cross-media 279
cross-subsidy 265
crosstalk 62–63
crypto 226, 232–5, 242
CSCW (Computer-Supported Cooperative Working) 322, 325–327
CTI (Computer-Telephony Integration) 151–2, 308
culture 281
customer service 15, 16
customers, business 9, 33

customers, domestic 9
customisation 321
cyberspace 3, 19

DARPA 178
data-glove 335
database 13, 16
datagram 45–46, 60
DCE (Distributed Computing Environment) 79
decency 26, 262, 285, 290–1
deception 25
defamation 26, 358
delay
   in ATM networks 52, 176, 345
   in computer networks 44, 45, 48, 353
   in CORBA architecture 80
   in satellite systems 129
   limits on 11, 17, 36, 67, 86, 327
   video processing 158, 371
delay-limited 59
Department of Defense 44, 334, 342
DES (Data Encryption Standard) 237–239, 243
design office, designer 327, 331
Desk Area Network 176
diagnosis, medical 2, 83
dialling, automatic 35
dialling, short-code 38, 142, 249
dialogue design 204
diamond scam 246
diffraction grating 110
digit, dialled 36
digital assistant 356
digital certificate 245, 249
digital signature 292
direct marketing 323
directory 38, 69, 78, 145, 320
disability 26, 262, 285, 288, 289, 341
disaster 353
Discrete Cosine Transform 370, 371
disintermediation 5, 318
disk 153–4, 174–5
dispersion 113
display 22
distribution 306
doctor 341

document reading room 320
documentation, electronic 12
Domain Name Server 74
drawing office 327–8
drawings, engineering 15
DSL (Digital Subscriber Loop) 351, 352
DSSS (Direct Sequence Spread Spectrum) 167–170
Dublin Core 184–185
duration, call 34
dynamic range 374–5
dynamic time warping 196–198

e-mail 11, 41, 44, 73, 287, 315, 319, 323, 326, 338
earth station 132
eavesdropping 230, 294
EDI (Electronic Data Interchange) 73, 306, 358
education 5, 15, 16, 27, 115, 179, 288, 298
eigenface 215
elderly 285, 289, 348
electromagnetic interference 280, 351
electronic
  banking 266
  cash 143, 258–9
  commerce 25, 224, 257–260, 322
  money 224, 257–260, 296–7, 303, 358
  purse 249, 258
  trading 9, 259, 306, 319, 321
  wallet 259
electronics, consumer 15, 18, 26, 138
emotion 325
empathy 326–7
employment 262, 297, 298
encoding
  differential 372
  digital 8
  speech 9, 124
  video 368–372
encryption
  applications 26
  chip 248
  GSM network 125

passive optical network 102
principles 225–226, 231–247, 296
set-top box 144, 273
endoscopy 83, 342
entertainment
  industry 4–5, 115, 140, 272, 282, 312–316
  technology 14–17, 22, 30, 66–7, 316, 351
EPG (Electronic Programme Guide) 280
epistemology 181
equality 285
error 94
Ethernet 43, 49, 166–7, 170, 353
ethics 284–5
Euclidean distance 194
European Union 275–281, 286, 296
evanescent field 100–101
exchange
  local 33, 38, 85, 102, 269
  main 31
  manual 35
  number of 114, 355
  software 70
  to-exchange network 20, 85
experts 15
extension 142

F-ratio 251–2, 254
facsimile 7, 49
fade 219
fairground worker 288
false rejection 255
fanzine 315
fares 4
fax 8, 274–5
FCC 263
FDM (Frequency Division Multiplex) 50, 99–100, 132
feature 214, 215, 220
fee, performance 283
Feistel cipher 236–7, 245
FHSS (Frequency Hopping Spread Spectrum) 167–170
file transfer 44
film 83, 290, 313
filter 105, 106, 367

finance 78
firewall 230–231
fleet management 287
foetal scan 342
follow-me 117
forgery 244
formant 254
Fourier, Jean-Baptiste 366, 370
frame (problem) 24, 188, 217
fraud 215, 223
free speech 285
freedom 298
frequency 364–6
frequency analysis 366, 369
frequency hopping 295
frequency-agile transmission 22
fuel cell 155
fulfilment 307, 310, 313, 318
fundamental (frequency) 366

games 216, 272, 276, 277, 314, 315, 316, 317
General Magic 69
globalisation 321
GLOBALSTAR 135
government 22, 262, 299
GPS (Global Positioning Satellite) 289, 304
graphics 15, 142
Green Paper 275–7, 280
greyhound racing 315
groceries 312
ground station 290
groupware 324
GSM 118, 124, 139, 152, 249, 287, 295
guidance, vehicle 305

hacker 72, 140, 226–7
handover/hand-off 121
handset 8
handwriting 26, 182
harmonic 366
hash function 245, 292
health 27
healthcare 78, 83–84, 179, 298, 332, 341, 345, 348
helpdesk 6, 41, 287, 299
hidden layer 213

hidden Markov model 196, 198–202
Highlands (of Scotland) 286
holiday 312
Hollywood 313
hologram 172
home computer *see also*, 'computer, personal' 357
home platform 356
home shopping *see also* 'shopping', 'retailing', 'teleshopping' 4, 63, 127–8, 136, 144–5, 148, 205, 308–9, 311, 312
home-bus 348
homeworking 287
HTML (Hypertext Mark-up Language) 185, 329
hybrid fibre-coax 63, 351
HyperLAN 170
hyperlink 289

ICE (Information, Communication, Entertainment) 11, 13, 272, 275, 279, 281, 350
ICO (satellite) 135
identity (proof of) 244
IDL (Interface Definition Language) 76–77, 80
image 15, 117, 177–9, 182, 219, 289, 292
  interpretation 209–210
  ownership of 283
image quality 309
image recognition 358, 210–219, 303
image transmission 334
image, TV 370, 67, 144
income, disposable 9
India 299
Information 1, 23
  crisis 177
  definition of 180
  networked 177
  services 82
  society 2, 27
  Superhighway 3
innovation 263
instant gratification 308
integrity 225

Intel 21
intellectual property 278
intelligent network 12, 13, 35, 37, 38, 40, 64, 65, 69, 80, 152, 355
intelligent peripheral 40
interactive TV 265, 314, 356–7
interactive voice response 205
interactivity 15, 277
interception 295
interface, stream 80
Interferometer, Fabry-Perot 103
Internet
 computing information network 66–7, 183
 example of distributed 74
 integration with telephony 64–6, 152, 355
 market and regulation 18–19, 276–7, 279, 285, 298
 name server 355
 network requirements 17, 44–6, 60, 63–4, 130
 Policy Registration Authority 247
 security issues 223, 238, 257
 shopping 308–9, 312
 teleworking 288, 327
Intranet 297
IP(Internet Protocol) 46, 51, 60
IRIDIUM 135
ISDN (Integrated Services Digital Network) 9, 158, 286, 310–311, 324, 326, 341–2, 351, 352
ISO/RM-OPD 80
isochronous (flow) 80
itinerant farmworker 288
IXC (Inter-exchange Carrier) 126

Jack (modelling system) 335
jamming 304
JAVA 68–69
jitter 110
joint 110

key 225, 232
key distribution centre 246
key management 232
key-escrow 296
key-ring 246

keyboard 22
keystream 235
keyword 184, 186, 219
kiosk, multimedia 308–311

labelling service 291–293
LAN (Local Area Network) 325, 327
 frequency hopping 295
 integration with telephony 150–2
 optical 153, 170–3
 radio 153, 167–170
 technical principles 42–44, 46, 49, 165–173, 353–4
landmark 304
landmarking 210–211
language 183
larynx 192
laser 20, 92, 95, 97–100, 103, 106, 111, 229
latency 80, 130–131
law 281, 290, 296
LEC (Local Exchange Carrier) 126
legislation 281
leisure 5
lens, lenticular 160
LEO (Low Earth Orbit Satellite) *see also* 'satellite' 119, 131–137
liberalisation 9, 271
library 23
linear-feedback shift register 235
link, computer 7
lithium niobate 110
lithography 101, 11460, 62, 87, 114
litigation 281
local call 277
local loop 264, 267, 352, 355
local rate 269
loss, optical fibre 93
loyalty (customer) 321

machine tool 328
machine translation 188–191
magic-box 227
mail order 4, 223
mainframe 165
maintenance 328, 335
MAN (Metropolitan Area Network) 107, 115

manufacturing 328, 332, 357
map, digital 290, 304
mark-up 23
McNair 18
meaning 180
mediation 320
medicine *see also* 'telemedicine' 177, 223, 348, 357
memory 154, 174
merger 277
messaging 117
messaging, voice 11
meta-data 184–185
meta-tag 184, 289, 292, 320, 329, 358
metering 277
microbending 228
microcell 167
microlens 110, 173
microsurgery 342
military 179
MIT 291
mobile code 67–69
mobile switching centre 121, 123
mobile telecommunications 280
mobile unit 121
mobility, universal personal 117, 124
modem 9, 33, 271, 275, 351
modem, cable 61, 63, 312, 325
modem, dial-back 226–227
modem, fax 50
modulation 31, 90
monopoly 279
Moore's Law 20, 21, 23, 149, 174, 360
motoring 4
mouse 22
movie 23, 63, 148, 314
MPEG 128
multimedia 11, 30, 38, 49, 60
  classification 219
  CORBA/TINA support for market and regulation 79–81, 276, 278, 327, 351
  mobile 22, 356
  radio 126
  rights 281–5
  satellite 132, 136

technical requirements 16–17, 44, 48, 316, 325, 353
terminal 140, 143
multimodal (transport) 301
multiplexing/multiplexor 8, 34, 50, 52, 54, 87, 114
  electrical 98
  frequency division 86
  optical 98, 109
  statistical 52–3, 55, 87, 88
  time division 86, 87
multipoint 49, 55, 81
multipoint conference bridge ('multi-party bridge') 157, 326
music 313, 314

name, call by 37
naming (CORBA) 78
nanotechnology 360
nation-state 26, 279, 285, 297
National Security Agency 238
negotiation 338
Network Information Centre 46
network
  all optical 20, 105
  cable 265
  campus 46
  cellular 126
  computer 16, 41, 44, 45, 49, 51, 58
  copper 29
  fixed 11, 29, 30, 117, 121, 123–125, 269
  global 2
  information 1, 15, 25, 66
  intelligent *see* 'intelligent network'
  inter-exchange 20
  management 67, 70, 72, 80, 265, 271
  mobile 9, 11, 37, 117, 356
  optical fibre 83–90, 91–94
  packet 33
  private 152
  provider 65
  satellite 126, 127, 129–137
  telecommunications 18, 51, 67
  telephone 34
  terrestrial 22, 127
  voice 37
Nineteen Eighty-four 294

NMR 83
noise 62, 89, 110, 168–9, 255, 374
nose, artificial 346
number/numbering
  personal 12, 37
  plan 269, 277
  portability 269–270
  range 269
  translation 39
numbers, Law of large 53

object (oriented) 75–7, 80, 356
obscenity 290
OFTEL 263, 272
oil-rigs 15
OMA (Object Management Architecture) 77–78
OMG (Object Management Group) 78–79, 80
on-demand 66, 83
  movies- 16, 127
  video- 282, 311–312, 313, 318
one-time pad 232–4
ontology 181, 331
Open Software Foundation 79
opinion poll 279
optical, optics
  amplifier 110–113
  coherent 112
  combiner 107, 109
  component 352
  coupler 101, 105, 107, 110, 111
  filter 92, 99, 103, 104
  free-space 171–3
  loss 110
  PON (Passive Optical Network) 101–103, 107, 231
  receiver 94, 109
  switching/routing 355
  tap 228–230
  transmitter 94–7, 109
  wavelength convertor 61, 112
  WDM (Wavelength Division Multiplexing) 20, 98, 99, 105, 173, 351, 354–5
optical fibre
  architecture 28, 60, 87–9
  cable 8, 228

  erbium-doped 20, 111–112
  LAN 153, 170–3
  local loop (*see also* 'optical PON') 61, 82
  potential 15, 19–20, 118, 286, 351–2
  principles 85–94, 97–103
  splicing 94
  splitter 92, 107–108
Oracle 351
ORB (Object Request Broker) 76–77
order-handling 308
organisation, virtual 73, 79
Orwell, George 294

PABX (Private Automatic Branch Exchange) 8, 38, 150–2
packet 34, 45–47, 48, 51, 130
pan 220
passcard 223
passive routing unit 107–109
password 226, 247, 248
pay-per-view 276, 280
payment 357
payment, secure 223
payphone 132, 143
payroll 41
PCS (Personal Communications Service) 124
persona 24, 316
personal organiser 66, 143
PET 83, 341
photodiode 92
PICS (Platform for Internet Content Selection) 290–3
picture grammar 210
PIN 223, 254, 259, 274
pipeline 334
piracy 281
pitch 192, 209, 254
plaintext 232, 238
point-of-sale 75, 307
point-to-multipoint 125
police 293
pop-up assistant 311
portable code *see* 'mobile code'
portable personality 249
postal services 4
Postmaster General 349

poverty 288
power, computing 20, 22
premium rate 273–274
price/pricing 21, 50, 277–8
prime number 239, 241
privacy 26, 102, 225, 262, 285, 293, 294, 295
private key 240–241, 245, 246, 248
product, consumer 30
product development 332
production 328
pronunciation 208–9
PTO (Public Telephone Operator) 265
pub 316
public key 228, 239–241, 247, 249, 296, 357
punch magnet 242

quality of service 11, 13, 44, 45, 47, 60, 67, 80–1, 263, 277, 339, 353
quantisation 363–4
quantum cryptography 229–230

radio (*see* chapter 4) 22, 23, 28, 29, 81, 89, 90
radio, cellular 21, 91, 295
rating service 291–293
recipe 318
recognition (identification)
  face 177, 178, 215, 359
  false 255
  fingerprint 26, 244, 250, 252–3, 254
  gesture 22, 358
  hand 253
  handwriting 250–2, 255
  image 358, 210–219, 303
  keyboard style 255
  pattern 161
  people 177
  retina 250, 256, 358
  speaker(verification) 253–5
  speech 22, 24, 38, 183, 191–204, 208, 221, 358
  voice 26, 155, 177, 253–4
refractive index 92–93, 104
region, less favoured 286
regulation 26, 262–281

regulator 22, 26, 262–281
regulatory boundary 26, 265, 266–267, 276, 278
reliability 49
replay attack 246
representation 298
retailer 306, 312, 318
retailing 18, 258, 289, 305–313, 337
retailing, wheel of 18
Revolution, Agrarian 297
Revolution, Industrial 297
rights, derived 283
ring 88
ring-tone 274
ringing 271
roaming 122
robot 342
robotics 161
route planner 301
router 42, 44, 46, 47, 51
routing 10, 35, 44, 122, 350
  least cost 64, 152
  time-of-day 40
  wavelength (*see also* 'optics, WDM') 106
royalty 282
RSA (Rivest, Shamir, Adleman) encryption 240–243
rule-based 182

safety 280
safety, optical system 172
sampling 362–3, 367
satellite *see also* 'LEO' 124, 141
  broadcasting 15, 17, 290, 299, 311
  digital 148
  geostationary 129, 136
  ground station 129
  link 316
  market and regulation 22, 118, 272–3
  radio 350–351
  transmission issues 87, 90–1
scattering 111
Science Museum 294
scrambling 235
screen, touch 310
screen, visual display 22, 155–7

SDH (Synchronous Digital Highway) 31, 32
security 25, 72, 79, 102, 223, 279, 295, 357
security policy 226
seduction 307–8
self-organisation 70, 360
self-service petrol station 349
sensor 22
server 16, 48, 63, 74
server, Web 74
service control point 39, 64, 65
service creation 40
service desk 6
service element 261
service, global 29
service management 40, 80
service provider 10, 11, 13, 63, 71, 261, 273, 274–5, 292
service switching point 39, 64
services, financial 5
services, information 67
services, public 289
set-top box 18, 63, 67, 128, 143–146, 148, 237, 272–3, 280
SGML (Standard Mark-up Language) 329
shop, electronic 312
shopping *see also* 'home shopping', 'retailing', 'teleshopping' 4, 16, 305, 307–313
shopping mall 303, 318
shrinkage 306
shuffling 236
sight difficulty 289
signalling 8, 9, 39, 64
signature 223, 244
signature, digital 244
simulation 335
singing telegram 68
SKYBRIDGE 136
small business 348
smartcard 143, 247–249, 253, 258, 301–3, 310, 317, 358
smell 161–2, 177
social fragmentation 314
social security 298
society 26, 348

solar energy 154–155
soliton 20, 112, 114
sound 15, 117, 163–5, 177, 219, 283
sound, voiced 192
speaker verification 253–5
speckle pattern 229
spectrum, optical 121
spectrum, radio 22, 94, 99, 118–119, 127, 280
spectrum, reuse 119
speech 10, 17, 22, 24, 89, 119178, 182, 364, 367
  encoding *see* 'encoding, speech'
  parts of 208–9
  production 191–192, 198–199
  recognition *see* 'recognition, speech'
  synthesis 22, 24, 191, 206–9, 216, 283, 289, 290
speed, processor 174
spoiling signal/spoiler 228, 284
sport 5
spy-in-the-cab 303
stagger-cast 128, 314
stakeholder 26, 261–299
standardisation 49, 67, 356
state *see* 'nation-state'
station (railway) 301
steganography 284
STEP 334
stethoscope 342
stop me and buy one 312
storage, disk
stream cipher 235
subscription 280
summarising 186
Sun Microsystems 69
supercomputer 20, 173
supermarket 307
supply chain 334
supply, power 153
surgery *see also* 'telesurgery' 2, 17, 84, 212, 223, 341, 345, 360
surveillance 9, 218, 294, 307
survivability 44–45
switch 42, 44, 86, 277
switch, circuit 51
switch, packet 51

INDEX

switching 20, 40, 60, 81, 85, 87, 114, 118, 350
switching (satellite) 132
switching, packet 35
SynchroLan 171
systems analyst 299

tachograph 303
tactile reader 289
tape, video 284
tapered star 31
tapping 228–230, 231, 242, 295
tariff 56, 116, 277
taste 177
tax gatherer 75–76
taxation 298
TCP (Internet Transmission Control Protocol) 47, 51, 60, 44, 54, 58, 130, 327, 345, 353–4
TDM (Time Division Multiplex) *see also* 'multiplexing' 31, 33, 34, 50, 98, 99–100, 104, 105, 132
teaching 287
tele-colonialism 2
telebanking 5, 143
Telecommunications Acts (UK) 282
TELEDESIC 131, 136
telemedicine *see also* 'medicine' 346
telemetry 9, 346
telephone/telephony 10, 22, 139, 350
    call model 29, 37, 50, 61–64, 80–1, 123
    cordless 125, 141–2
    domestic 7, 8, 9
    mobile 69, 81, 82, 118, 122, 141
    multimedia 66
    office 150
telepresence 22, 161
teleprinter 242, 294
telesales 151, 299, 339
teleshopping *see also* 'home shopping', 'retailing', 'shopping' 314
telesurgery *see also* 'surgery' 348
Teletext 351
television: *see* TV 16
teleworker 287
teleworking 149, 322–5, 352
telex 179, 182

terrestrial broadcast 311
terrestrial radio 350–351
terminal 10, 12, 35, 51, 64, 65, 66, 117, 138, 141–145, 259
    business 150
    consumer 15
    domestic 149
    dumb 41
    information 15
    intelligent
    mobile 37
    multimedia 40, 301
    public 266, 289
    sales 15
terrorist 304
testing 328
text 15, 178, 179, 219
text-to-speech synthesis 208
TF-IDF (Term Frequency-Inverse Document Frequency) 188, 189
theft 25, 283–4, 306
thermoelectricity 154
three-dimensions 20, 148, 159–60, 357
time-to-market 321
TINA (Telecommunications Networking Information Architecture) 25, 80, 356
token ring 43
toll 301, 302, 303, 358
tones, multifrequency 64
touch 161–3, 346
tour operator 319
towbar 303
trading (CORBA) 78, 79
traffic congestion 301
traffic report 318
train (railway) 301
training 287
transaction 308
translation, artificial 358
translation, number 37, 65, 65
transponder 302
transport 301–305, 348
transport, public 301, 305
travel 5, 7, 14
travel agent 318
travelogue 312
tree-and-branch network 88

trust 25, 26, 222–224, 247, 326, 336–7
trusted third party 246, 296
TV 23, 118, 307
  Business 337
  camera 17, 158, 213, 216, 252, 266, 311
  digital 15, 22, 108, 127–8, 139, 276, 280, 311
  high definition 128
  interactive 16, 128, 143
  market, regulation, convergence 127, 138–9, 273, 276–7, 290, 294, 314, 356
  requirements and technology 60–63, 66, 83, 89, 103, 107

ultrasound 83
Uncertainty Principle 230
unemployment 286
universal service 280, 285
URL 74
US Airforce 335
US Immigration Department 253

value added service 266–268
value chain 305
value, call by 37
VCR 139
verification 258
video 17, 24, 119, 127, 310
  conferencing 8, 9, 15, 17, 55, 65, 105, 124, 157, 319–320, 322, 332
  desktop 157–8
  on-demand *see* 'on-demand, video-'
  signal processing 158, 367–372
videophone 11, 15, 67, 140, 215, 287, 319, 322, 324, 350, 352
virtual
  art gallery 282
  business 38, 320–341, 336
  Christmas dinner 317
  community 317
  design 335
  enterprise 320–341
  office 328
  organisation 320–341, 359
  prototyping 331–2

reality 148, 212, 335
science park 319–320
virus 72
visualisation 332
vocal cords 192
voice circuit 6
voice response 308
voice, identification/recognition *see* 'recognition, speech/voice'
voice-mail 319
voiceprint 253–4
vowel 195
VPN (Virtual Private Network) 12, 65–6, 308
VSAT (Very-Small-Aperture Terminal) 287, 337

wages 285, 299
WAN (wide area network) 13, 44, 51
warehouse 287, 306
watermark 79, 284–5, 292
watershed 282
wave 89, 91
waveguide, helical 117
wavelength 89, 91
wealth 299
Web TV 277
weighting network 213
welfare 346
whiteboard 322
window, optical 97
wire 15, 28, 31, 35, 60, 6284, 85, 86, 89
wire, private 41
wireless fibre 126
wiring 142
work-groups 15
work-scheduling 338
workflow 338
workstation 41, 65, 153, 170, 175, 327
workstation, mobile 30
World Wide Web 44, 68, 184, 288, 290, 317, 31, 356

X-ray 341, 342
XML (Extensible Mark-up Language) 185, 330

Yellow Pages 69, 79

Books are to be returned on or before the last date below.

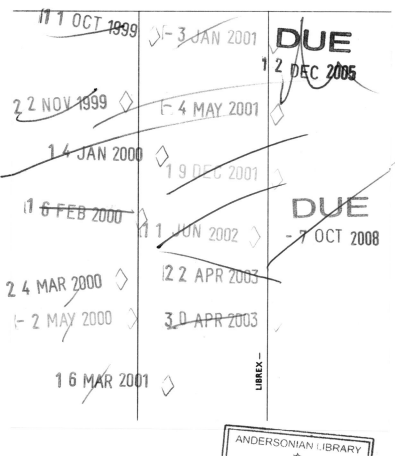